United States Nuclear Regulatory Commission

Protecting People and the Environment

NUREG-2110

I0493893

xLPR Pilot Study Report

Office of Nuclear Regulatory Research

AVAILABILITY OF REFERENCE MATERIALS
IN NRC PUBLICATIONS

NRC Reference Material

As of November 1999, you may electronically access NUREG-series publications and other NRC records at NRC's Public Electronic Reading Room at http://www.nrc.gov/reading-rm.html.
Publicly released records include, to name a few, NUREG-series publications; *Federal Register* notices; applicant, licensee, and vendor documents and correspondence; NRC correspondence and internal memoranda; bulletins and information notices; inspection and investigative reports; licensee event reports; and Commission papers and their attachments.

NRC publications in the NUREG series, NRC regulations, and *Title 10, Energy*, in the Code of *Federal Regulations* may also be purchased from one of these two sources.
1. The Superintendent of Documents
   U.S. Government Printing Office
   Mail Stop SSOP
   Washington, DC 20402–0001
   Internet: bookstore.gpo.gov
   Telephone: 202-512-1800
   Fax: 202-512-2250
2. The National Technical Information Service
   Springfield, VA 22161–0002
   www.ntis.gov
   1–800–553–6847 or, locally, 703–605–6000

A single copy of each NRC draft report for comment is available free, to the extent of supply, upon written request as follows:
Address: U.S. Nuclear Regulatory Commission
           Office of Administration
           Publications Branch
           Washington, DC 20555-0001
E-mail: DISTRIBUTION.SERVICES@NRC.GOV
Facsimile: 301–415–2289

Some publications in the NUREG series that are posted at NRC's Web site address http://www.nrc.gov/reading-rm/doc-collections/nuregs are updated periodically and may differ from the last printed version. Although references to material found on a Web site bear the date the material was accessed, the material available on the date cited may subsequently be removed from the site.

Non-NRC Reference Material

Documents available from public and special technical libraries include all open literature items, such as books, journal articles, and transactions, *Federal Register* notices, Federal and State legislation, and congressional reports. Such documents as theses, dissertations, foreign reports and translations, and non-NRC conference proceedings may be purchased from their sponsoring organization.

Copies of industry codes and standards used in a substantive manner in the NRC regulatory process are maintained at—
           The NRC Technical Library
           Two White Flint North
           11545 Rockville Pike
           Rockville, MD 20852–2738

These standards are available in the library for reference use by the public. Codes and standards are usually copyrighted and may be purchased from the originating organization or, if they are American National Standards, from—
           American National Standards Institute
           11 West 42nd Street
           New York, NY 10036–8002
           www.ansi.org
           212–642–4900

United States Nuclear Regulatory Commission

Protecting People and the Environment

NUREG-2110

xLPR Pilot Study Report

Manuscript Completed: May 2012
Date Published: May 2012

Prepared by:
David Rudland
Craig Harrington

NRC Project Manager:
David Rudland

Office of Nuclear Regulatory Research

CITATIONS

This report was prepared by:

U.S. Nuclear Regulatory Commission (NRC),
Office of Nuclear Regulatory Research (RES)
Washington, DC 20555-0001

Principal Investigator
D. Rudland

Electric Power Research Institute (EPRI)
3420 Hillview Avenue
Palo Alto, CA 94304

Principal Investigator
C. Harrington

This report describes research sponsored jointly by the U.S. Nuclear Regulatory Commission (NRC), Office of Nuclear Regulatory Research (RES) and the Electric Power Research Institute (EPRI).

This report is a corporate document that should be cited in the literature in the following manner:

xLPR Pilot Study Report. U.S. NRC-RES, Washington, DC, and EPRI, Palo Alto, CA: NUREG-2110 and EPRI 1022860. 2012.

ABSTRACT

Under the auspices of an addendum to the memorandum of understanding between the Electric Power Research Institute and the U.S. Nuclear Regulatory Commission's Office of Nuclear Regulatory Research for cooperative research, a pilot study has been completed to evaluate the feasibility of developing a fully probabilistic, fracture-mechanics-based computational tool to evaluate the rupture probability of reactor coolant piping. This project, known as xLPR for Extremely Low Probability of Rupture, is initially focused on evaluating pipe rupture probabilities within Alloy 82/182 dissimilar metal welds located in lines licensed for leak-before-break (LBB) as allowed under General Design Criterion 4, "Environmental and Dynamic Effects Design Bases," of Appendix A, "General Design Criteria for Nuclear Power Plants," to Title 10 of the *Code of Federal Regulations* Part 50, "Domestic Licensing of Production and Utilization Facilities." The current LBB regulatory basis does not allow for assessment of piping systems subject to active degradation mechanisms, such as primary water stress-corrosion cracking, which has been detected in some systems that have been granted LBB approval. Although the piping systems susceptible to this type of corrosion have been shown through deterministic arguments to comply with the regulations, no fully probabilistic tool currently exists to directly assess this compliance.

Development of such a complex probabilistic computational tool is a daunting technical challenge, and the project team determined that a pilot study was necessary. The three key goals addressed include establishing the fundamental feasibility of such an undertaking, assessing whether the proposed organizational approach could accomplish the task, and informing the decision on the most appropriate computational platform to employ.

This report summarizes the results of that pilot study and provides an overview of the complete project documentation contained in a number of more detailed reports. The xLPR Pilot Study team demonstrated that it is feasible to develop a modular-based computer code for the determination of probability of rupture for LBB-approved piping systems. Furthermore, while the organization established to manage the project and accomplish the technical work of the Pilot Study successfully met that challenge, it identified important improvement opportunities that will be addressed as the project moves forward. Finally, substantial knowledge and experience were gained through the development of two parallel Pilot Study computational codes using a commercially licensed simulation framework code in one case and an open source framework code in the other. This approach has provided a strong basis for computational platform selection for further xLPR development. The xLPR Pilot Study successfully met its three key objectives and has established a solid base of knowledge and experience supporting further development.

Keywords
xLPR
LBB
Pipe rupture

CONTENTS

LIST OF FIGURES

LIST OF TABLES

EXECUTIVE SUMMARY

Title 10 of the *Code of Federal Regulations* Part 50, "Domestic Licensing of Production and Utilization Facilities," Appendix A, "General Design Criteria for Nuclear Power Plants," General Design Criterion (GDC) 4, "Environmental and Dynamic Effects Design Bases," states, in part, that the dynamic effects associated with postulated reactor coolant system pipe ruptures may be excluded from the design basis when analyses reviewed and approved by the U.S. Nuclear Regulatory Commission (NRC) demonstrate that the probability of fluid system piping rupture is extremely low under conditions consistent with the design basis. NUREG-0800, "Standard Review Plan for the Review of Safety Analysis Reports for Nuclear Power Plants: LWR Edition" (also known as the SRP), Section 3.6.3, "Leak-Before-Break Evaluation Procedures," describes leak-before-break (LBB) deterministic assessment procedures that have been used to date to demonstrate compliance with the GDC 4 requirement. Currently, SRP Section 3.6.3 does not allow for assessment of piping systems with active degradation mechanisms, such as primary water stress-corrosion cracking (PWSCC), which is currently occurring in systems that have been granted LBB exemptions. Even though the piping systems experiencing PWSCC have been shown through qualitative arguments to comply with the regulations, no tool currently exists to quantitatively assess compliance with this criterion.

From the NRC staff's perspective, a long-term goal is to develop a modular-based probabilistic fracture mechanics (PFM) tool capable of determining the probability of failure for reactor coolant system components. The need for this modular-based code is strongly driven by the need to quantitatively assess the LBB-approved piping system's compliance with GDC 4 on an interval (time) basis. To meet this need, the NRC Office of Nuclear Regulatory Research has entered into a cooperative program with the Electric Power Research Institute (EPRI) through an addendum to the memorandum of understanding to define, design, and develop this code, which will be comprehensive with respect to known challenges, vetted with respect to scientific adequacy of models and inputs, flexible enough to permit analysis of a variety of inservice situations, and sufficiently adaptable to accommodate evolving and improving knowledge. The code has a modular structure so that as additional situations arise, additions or modifications can be easily incorporated without code restructuring. Based on the terminology of GDC 4, this program and code are titled Extremely Low Probability of Rupture (xLPR).

Under the addendum to the memorandum of understanding, a project management organizational structure with balanced NRC and industry representation was developed consisting of four topical technical task groups coordinated by an overarching Project Integration Board. A diverse team of experts in the various technical specialties involved in this complex analytical project was assembled under funding from both the NRC and EPRI to develop the detailed analytical methodologies, identify the necessary input data and mathematical models, and assemble them into a functioning set of computational tools. The technical task groups and highlights of their areas of responsibility are as follows:

- Computational Task Group: integration of the computational elements (models) into a robust, fully developed, tested, and verified computational tool

- Models Task Group: selection, documentation, and coding of the mathematical model building blocks

- Inputs Task Group: identification, collection, and presentation of the data required for the models and the sample problem as input tables and distributions

- Acceptance Criteria Task Group: formulation of probabilistic acceptance criteria for assessing the code results

The development of a sophisticated probabilistic software tool that meets quality assurance (QA) and technical requirements is a technically and programmatically challenging task. The management structure, the probabilistic framework, and data handling are just a few of the issues that must be addressed early in the software development effort. To meet this need, a pilot study was conducted. This pilot study is a proof-of-concept effort to develop an initial assessment tool for dissimilar metal pressurizer surge nozzle welds, for which considerable publicly available information exists. The xLPR Pilot Study objective was to demonstrate the overall feasibility of the proposed probabilistic code, exercise the intended cooperative organizational model, and determine the computational framework structure that is most appropriate to meet the longer term program goal of developing a modular-based, PFM tool capable of determining the probability of failure for reactor coolant system components. This report presents an overall summary of the Pilot Study, the xLPR Version 1.0 codes, and the results obtained.

To assess the capabilities of computational framework architecture, two unique framework codes were developed in the Pilot Study to investigate the advantages and disadvantages of the different approaches used by each code. GoldSim was used to develop a commercial software version of the xLPR Model, and an open source software version, Structural Integrity Assessment Modular—Probabilistic Fracture Mechanics (SIAM-PFM), was also developed.

As part of the Pilot Study, the two xLPR codes were developed following a strict configuration management (CM) structure. The xLPR CM structure consists of a systematic approach applied to both the developed software and models to ensure that some of the basic fundamentals of a QA program are met, including (1) access control, (2) version control, (3) verification (e.g., checking), and (4) traceability (e.g., documentation). The xLPR CM approach included documentation of each step in the process. The study team implemented the CM process as detailed in a series of guidance documents that outline the specific steps for each of four key components of the xLPR pilot program: (1) module development, (2) framework development, (3) model parameters and inputs for the Pilot Study test case, and (4) xLPR model production runs and uncertainty/sensitivity analyses for the Pilot Study test case. Even though code development followed a strict CM plan, that CM plan was not linked to a formal QA process, which will occur in later versions of the code.

The flow of the xLPR Version 1.0 code is centered on a time-based history of events in which PWSCC initiates flaws that grow until failure. Subject matter experts within the Models Group, together with members of the Computational Group, developed the technical basis for this behavior by using the predefined CM process to code, compile, and verify the modules needed for this purpose. These modules included loads with weld residual stress, crack initiation, crack growth, crack coalescence, crack stability, crack opening displacement, leakage, inspection, and mitigation. Both the commercial and open source applications used these self-contained modules, which were linked to create two xLPR framework codes that control the time flow of the analyses and properly account for and propagate the problem uncertainties.

The xLPR framework for calculating the probability of primary system pipe rupture invokes a systematic approach to uncertainty characterization and the propagation of probability distributions. To better understand the effects of uncertainty on the distribution of the output parameters desired, the uncertainty is classified as aleatory (random or irreducible) or epistemic (resulting from a lack of knowledge or reducible). The uncertainty in the input is propagated through the model by use of sampling-based methods. The appropriate way to propagate uncertainty ultimately depends on the computational constraints, as well as the nature of the inputs and outputs under consideration. Within the xLPR Pilot Study, several sampling methods were evaluated, such as simple random sampling, Latin hypercube sampling, and discrete probability distribution space sampling. In addition, importance sampling was used to estimate the low-probability events.

A sample problem statement was developed to demonstrate the feasibility of conducting analyses to assess the probability of rupture in dissimilar metal welds on the pressurizer surge nozzle and to compare the results from the two frameworks developed in this effort. This problem statement consisted of two deterministic analyses, a probabilistic base case, and a series of sensitivity analyses to demonstrate the features of the Version 1.0 code, as well as to provide a proof of concept for the overall xLPR methodology. These runs focused on the demonstration of the calculation of probability of rupture with and without mitigation, inspection, and leak detection, while accounting for epistemic and aleatory uncertainties.

Not unexpectedly, the probabilistic base case, which contained high weld residual stress with no inspection, mitigation, or leak detection, produced relatively high mean probabilities of rupture using both the GoldSim and Structural Integrity Assessment Modular (SIAM) frameworks. In fact, the two codes gave approximately the same values, with the difference attributed to the handling of the crack initiation model. Further analyses suggest that the calculated results are stable and highly driven by the epistemic uncertainty. Because of the high rate of crack growth and assumptions made about the crack behavior, the benefit of leak detection and inspection was not as large as expected. Parameter sensitivity studies conducted on the base case suggested that the weld residual stress and the crack initiation parameters (all characterized as epistemic) were controlling the uncertainty in the probability of rupture.

The sensitivity studies investigated the effects of stress mitigation, chemical mitigation, crack initiation model, and weld residual stress on the probability of rupture as compared to the base case. The conclusions from these sensitivity studies indicate the following:

- For stress mitigation, the effects are seen in both the crack initiation and growth models. The mitigation was applied at 10, 20, or 40 years. The results from these analyses demonstrate that the effective application of the assumed stress-based mitigation could cause the probability of rupture to no longer increase with operating time after the mitigation begins.

- Two chemical mitigation cases changed the hydrogen content in the water from the base case value of 25 cubic centimeters per kilogram (cc/kg) to 50 cc/kg and 80 cc/kg. As expected, additional hydrogen decreases the probability of rupture at 60 years by about 50 percent for 50 cc/kg but has little additional effect for 80 cc/kg.

- Leak detection over the range of 1 to 10 gallons per minute reduced the rupture probability at 60 years by a factor of 10.

- Inservice inspection every 2 years reduced the 60-year rupture probability by a factor of 70, while inservice inspection every 10 years reduced it by only a factor of 2.

- Changing the crack initiation model had very little impact on the overall rupture probability. This was expected since each of the initiation models is empirically based and calibrated to the same service history.

- Changing the crack initiation parameters from epistemic to aleatory had a large effect on the probability of rupture distribution. While the mean value stays the same, when the uncertainty in initiation time is characterized as epistemic, there is a 0 percent chance of any future rupture 50 percent of the time. When the uncertainty is characterized as aleatory, there is a 35 percent chance of any future rupture 50 percent of the time.

- Changing the weld residual stress had a large impact on the rupture probability. Changing the residual stress from a nozzle geometry without a safe end weld to one with a safe end weld decreased the rupture probabilities at 60 years by 2 orders of magnitude.

In many of the cases with low residual stress and/or inspection and leak detection, not enough samples were taken to produce stable results. Therefore, importance sampling was needed. In all cases considered, the weld residual stress and an initiation parameter (B1) were importance sampled. The results indicate that with inspection and leak detection, probabilities down to 10^{-6} at 60 years can be calculated with reasonable confidence. However, when mitigation is added, probabilities down to 10^{-9} at 60 years are calculated, but the confidence in the mean values is very poor. Additional realizations (predictions of rupture) are required to increase the confidence in these results.

During this investigation, the xLPR project team developed an appreciation for the complexity of this problem, and the structure needed for successful completion of a comprehensive PFM code. Through the process, the team learned many important lessons. These include not only technical lessons from the module and framework development and implementation, but also lessons related to organization and program management, such as the following:

- Three very important organizational structure aspects are required for program success:
  - dedicated team members, whose qualifications cover the important aspects of the group responsibility
  - an enthusiastic team and group leadership
  - an efficient communication process within and among the teams

- CM is only a small part of QA. Establishing a program QA plan and controls for xLPR is the essential first step in the continuing development process. The xLPR program needs to have a transparent and traceable CM system that will cover the xLPR code life cycle.

- A well-written, unambiguous software requirements document must be developed and followed for future xLPR versions.

- From the models standpoint, certain assumptions were made because of the limited scope of the Pilot Study. Model-based limitations and lessons learned include the following:

- Manufacturing defects and fatigue initiation and growth were ignored in the Pilot Study. Both should be included in future versions, since their omission may lead to nonconservative rupture probabilities.

- The load module should be updated to include a more realistic weld residual stress model and transient definitions.

- Considering only circumferential cracks may overpredict the rupture probabilities. The addition of axial cracks may reduce the rupture probabilities because of their higher leakage probabilities which would lead to early repair or mitigation.

- Assuming idealized flaw shapes and simplistic transitions from a surface crack to a through-wall crack may cause an overestimate of the leak rate.

- More realistic surface crack stability, inspection, and mitigation models are required for making best estimate predictions of their effects.

- Emphasis should be placed on efficient data storage, data handling, and postprocessing to improve the running of the code.

- Importance sampling is necessary for the calculation of the probability of rupture in piping systems. Processes and procedures for identifying the variables that need to be importance sampled should be emphasized. Adaptive sampling or other reliability methods should be considered.

- The classification of uncertainty is very important to understanding the overall uncertainty in the probability of rupture. Consideration of uncertainty is critical at all levels of development of a complex system. Knowledge of which variables control the rupture and which part of the uncertainty in those variables is epistemic and can be reduced will not only inform the regulators, but will also help direct future research in this area. In other words, xLPR can be used to prioritize research efforts and degradation management strategies to quantitatively improve safety.

Finally, based on an independent comparison between GoldSim and SIAM, a cost analysis, and the long-term prospects of the software, the xLPR project team recommends that the future versions of xLPR be developed using the GoldSim commercial software as the computational framework. Also, the complete xLPR Pilot Study effort, which includes not only the code development efforts, but the management structure, the pilot statement problem, and the detailed analysis of the results, demonstrates that it is feasible to develop a modular-based computer code for the determination of probability of rupture for LBB approved piping systems.

ACKNOWLEDGMENTS

The development of the Version 1.0 xLPR code was a group effort involving a variety of experts across many fields of expertise from the NRC, EPRI, and their contractors. The success of the program reflects the dedication of the xLPR team, the strength of its leadership, and the generous support from both the NRC and EPRI. There are many people to thank, including members from the Computational, Models, Inputs, and Acceptance Criteria Groups, as well as the Project Integration Board. Every person on this team made valuable contributions, and their efforts are sincerely appreciated.

Project Integration Board

Craig Harrington—EPRI

Aladar Csontos—NRC

Robert Hardies—NRC

Denny Weakland—Ironwood Consulting

David Rudland—NRC

Bruce Bishop—Westinghouse Electric Co. LLC (WEC)

Eric Focht—NRC

Guy DeBoo—Exelon

Marjorie Erickson—Phoenix Engineering Associates, Inc. (PEAI)

Gary Stevens—NRC

Howard Rathbun—NRC

Mark Kirk—NRC

Glenn White—Dominion Engineering, Inc. (DEI)

Computational Group

David Rudland—NRC

Bruce Bishop—WEC

Nathan Palm—WEC

Patrick Mattie—Sandia National Laboratories (SNL)

Cedric Sallaberry—SNL

Don Kalinich—SNL

Jon Helton—SNL

Hilda Klasky—Oak Ridge National Laboratory (ORNL)

Paul Williams—ORNL

Robert Kurth—Engineering Mechanics Corporation of Columbus (EMC^2)

Scott Sanborn—Pacific Northwest National Laboratory (PNNL)

David Harris—Structural Integrity Associates (SIA)

Dilip Dedhia—SIA
Anitha Gubbi—SIA

Models Group
Marjorie Erickson—PEAI
Gary Stevens—NRC
Howard Rathbun—NRC
David Rudland—NRC
John Broussard—DEI
Glenn White—DEI
Do-Jun Shim—EMC^2
Gery Wilkowski—EMC^2
Bud Brust—EMC^2
Cliff Lange—SIA
Dave Harris—SIA
Steve Fyfitch—AREVA NP Inc.
Ashok Nana—AREVA NP Inc.
Rick Olson—Battelle
Darrell Paul—Battelle
Lee Fredette—Battelle
Craig Harrington—EPRI
Gabriel Ilevbare—EPRI
Frank Ammirato—EPRI
Patrick Heasler—PNNL
Bruce Bishop—WEC

Inputs Group
Eric Focht—NRC
Mark Kirk—NRC
Guy DeBoo—Exelon
Paul Scott—Battelle
Ashok Nana—AREVA NP Inc.
John Broussard—DEI
Nathan Palm—WEC
Pat Heasler—PNNL
Gery Wilkowski—EMC^2

Acceptance Criteria Group
Mark Kirk—NRC
Glenn White—DEI
Aladar Csontos—NRC
Robert Hardies—NRC
David Rudland— NRC
Bruce Bishop—WEC
Robert Tregoning—NRC

ACRONYMS AND NOMENCLATURE

ADAMS	Agencywide Documents Access and Management System
ASME	American Society of Mechanical Engineers
BWR	boiling-water reactor
cc/kg	cubic centimeters/kilogram (unit of concentration)
CM	configuration management
CNWRA	Center for Nuclear Waste Regulatory Analyses
COD	crack opening displacement
DEI	Dominion Engineering Inc.
EMC^2	Engineering Mechanics Corporation of Columbus
EPRI	Electric Power Research Institute
GDC	general design criterion/criteria
gpm	gallons per minute
GTG	GoldSim Technology Group
GUI	graphical user interface
II	inspection interval
ISO	International Organization for Standardization
LBB	leak-before-break
LD	leak detection
NQA	Nuclear Quality Assurance
MRP	Materials Reliability Program (EPRI)
NRC	U.S. Nuclear Regulatory Commission
NUREG	NRC technical report designation
ORNL	Oak Ridge National Laboratory
PEAI	Phoenix Engineering Associates, Inc.
PDF	probability density function
PFM	probabilistic fracture mechanics
PIB	Project Integration Board
PNNL	Pacific Northwest National Laboratory
POD	probability of detection
PWR	pressurized-water reactor
PWSCC	primary water stress-corrosion cracking
QA	quality assurance
RES	Office of Nuclear Regulatory Research
SC	surface crack
SIA	Structural Integrity Associates

SIAM	Structural Integrity Assessment Modular
SIAM-PFM	Structural Integrity Assessment Modular—Probabilistic Fracture Mechanics
SNL	Sandia National Laboratories
SQA	software quality assurance
SRP	Standard Review Plan
STP	standard temperature and pressure
TWC	through-wall crack
V1.0	xLPR Pilot Study Version 1.0 codes
V2.0	xLPR Version 2.0—initial production code
WEC	Westinghouse Electric Co. LLC
xLPR	extremely low probability of rupture
yr	year

1
INTRODUCTION

1.1   xLPR—Background and Motivation

Within the U.S. nuclear regulatory framework, the general design criteria (GDC) contained in Title 10 of the *Code of Federal Regulations* Part 50, "Domestic Licensing of Production and Utilization Facilities," Appendix A, "General Design Criteria for Nuclear Power Plants," are the cornerstones establishing basic design requirements that nuclear power facilities in the United States must meet. GDC 4, "Environmental and Dynamic Effects Design Bases," presents specific compliance challenges to the pressurized-water reactor (PWR) fleet with the requirement to protect against the local dynamic effects of pipe ruptures. These issues were generally resolved through a deterministic, analytical approach that has come to be known simply as leak-before-break (LBB). GDC 4 allows LBB through an addition to the text stating "dynamic effects associated with postulated pipe ruptures in nuclear power units may be excluded from the design basis when analyses reviewed and approved by the Commission demonstrate that the probability of fluid system piping rupture is extremely low under conditions consistent with the design basis for the piping." NUREG-0800, "Standard Review Plan for the Review of Safety Analysis Reports for Nuclear Power Plants: LWR Edition" (also known as the SRP), Section 3.6.3, "Leak-Before-Break Evaluation Procedures," describes deterministic LBB assessment methodologies that are acceptable to the staff of the U.S. Nuclear Regulatory Commission (NRC). SRP Section 3.6.3 also includes a key condition that the piping system not be subject to any known active degradation mechanisms. Exclusion of active degradation mechanisms is both a conservative simplifying assumption and an artifact of prevailing analytical capability limitations in the 1980s when these regulations were put in place.

LBB has generally been applied to select portions of the reactor coolant system piping in domestic PWRs. However, when the revision to GDC 4 was promulgated, intergranular stress corrosion cracking had already been identified as an active degradation mechanism of concern to boiling-water reactors (BWRs), and thus LBB has not been implemented within the BWR fleet. The more recent identification of primary water stress corrosion cracking (PWSCC) in PWR locations previously approved for LBB raises questions that now must be resolved. While industry and the NRC have taken appropriate actions to adequately ensure fleet safety relative to PWSCC for the intermediate term, a long term technical resolution is needed.

The prescribed LBB analytical methodologies are deterministic approaches to address a fundamentally probabilistic design requirement. Although the LBB technical basis is sound, the linkage between the deterministic analytical methodology and the probabilistic design criteria is not sufficiently robust to allow direct incorporation of rigorous analytical treatment of active degradation mechanism effects. The Electric Power Research Institute (EPRI) Materials Reliability Program (MRP) and the NRC Office of Nuclear Regulatory Research (RES) have therefore initiated a cooperative effort to take advantage of advances in analytical methods and computational capabilities to develop a new, more robust technical basis and analytical methodology to demonstrate and assess compliance with the "extremely low probability of rupture" standard. A project-specific addendum [1] to the general memorandum of understanding for such cooperative research activities between the NRC and EPRI formally established this cooperative effort. Although initially focused on resolving the PWSCC challenge for PWRs, the intent of the Extremely Low Probability of Rupture (xLPR) project is to develop a fully probabilistic approach applicable to a range of active degradation mechanisms associated with both BWRs and PWRs. The resulting computer code will be comprehensive with respect to known materials degradation challenges, vetted with respect to the scientific adequacy of models and inputs, flexible enough to permit analysis of a variety of in-service situations, and sufficiently adaptable to accommodate evolving and improving knowledge.

1.2   Purpose of This Report

Development of a sophisticated probabilistic software tool that meets necessary technical requirements and incorporates relevant quality assurance (QA) needs is a technically and programmatically challenging task. Developing the project management structure, defining the probabilistic framework, modeling complex physical phenomena, and collecting and handling data are just a few of the issues that must be addressed very early in the software development effort. Given this inherent complexity, a pilot study was undertaken with limited goals of (1) demonstrating the feasibility of the concept, (2) informing key computational platform decisions, and (3) exercising the process approach proposed for developing a computational tool to evaluate the probability of degradation in piping systems leading to rupture.

This report presents a high-level summary of the entire effort that constitutes the xLPR Pilot Study conducted from spring 2009 through fall 2010 and addresses project outcomes and recommendations for each of these three major Pilot Study goals. To provide context for the discussion of outcomes, the report begins with a discussion of the goals and an overview of the two Pilot Study codes and their development process. The report then presents Pilot Study outcomes and recommendations with respect to each of these goals and with relevance to subsequent phases in the development of xLPR, followed by an overview of key lessons learned.

Separate reports have been prepared to document the many important aspects of the Pilot Study in greater detail than will be presented here; they provide the essential foundation of this summary report. Table 1-1 presents the complete list of reports, and Figure 1-1 depicts the report hierarchy. All reports are publicly available through either the NRC Agencywide Documents Access and Management System (ADAMS) Web site or the EPRI public Web site.

Table 1-1: xLPR Pilot Study Reports

Report Title	Developer	Identifier
xLPR Pilot Study Final Report	NRC & EPRI	NUREG-2110 EPRI PID 1022860
GSxLPR and SIAMxLPR Comparison Report [7]	CNWRA (Southwest Research Institute)	ML111510924
xLPR Version 1.0 Report, Technical Basis and Pilot Study Problem Results [5]	xLPR Computational Group	ML110660292
xLPR Framework (GoldSim) Model User's Guide [8]	Sandia National Laboratory	ML110700017 SAND2010-7131
Structural Integrity Assessments Modular—Probabilistic Fracture Mechanics (SIAM-PFM): User's Guide for xLPR [9]	Oak Ridge National Laboratory	ML110700023
Development, Analysis, and Evaluation of a Commercial Software Framework for the Study of Extremely Low Probability of Rupture (xLPR) Events at Nuclear Power Plants [2]	Sandia National Laboratory	ML110700019 SAND2010-8480
SIAM-xLPR Version 1.0 Framework Report [3]	Oak Ridge National Laboratory	ML110700026
Models and Inputs Selected for Use in the xLPR Pilot Study [4]	xLPR Models/Input Groups	EPRI PID 1022528

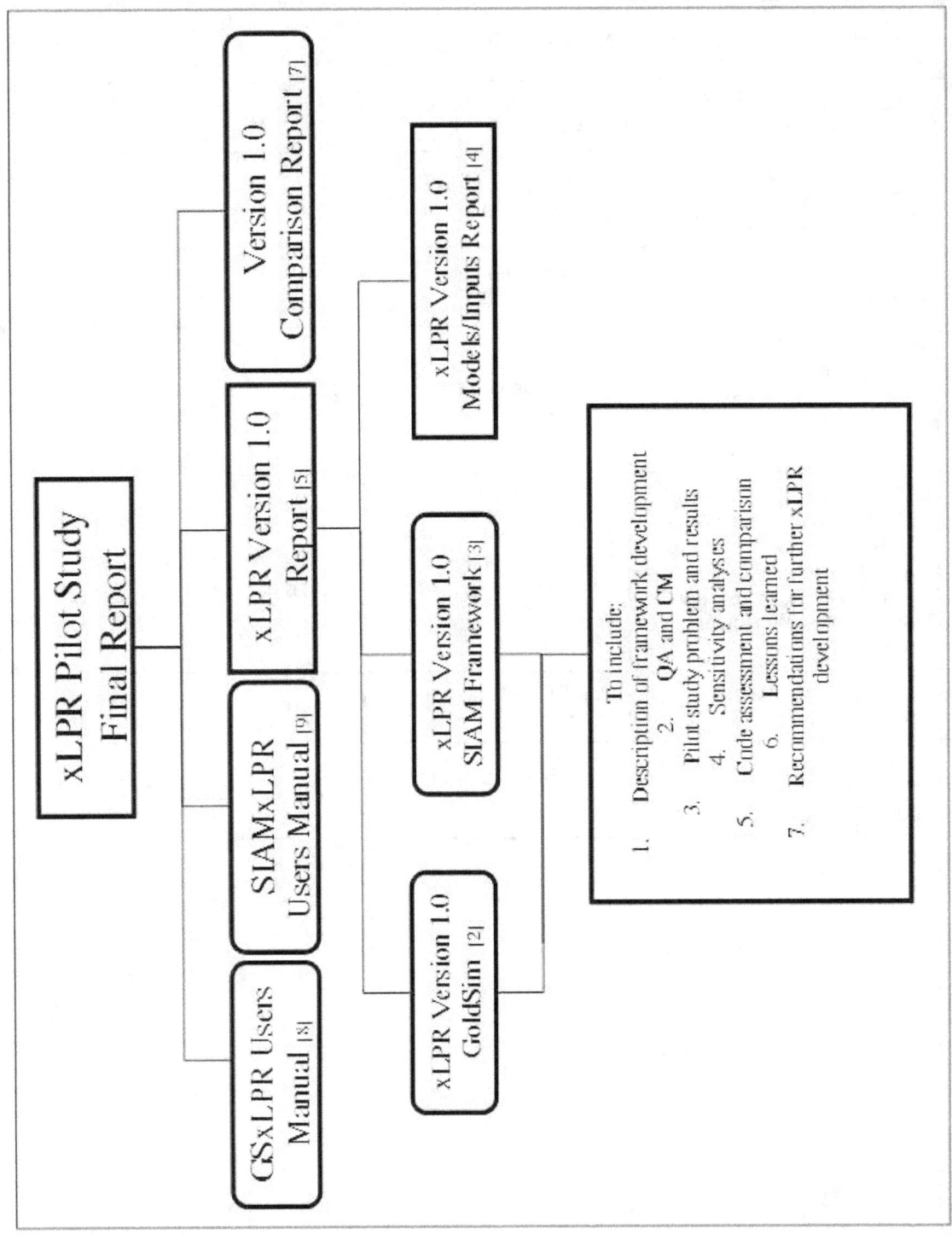

Figure 1-1 xLPR Pilot Study report chart

2
PILOT STUDY DESCRIPTION

The complexity of the broader xLPR project goal overwhelmed initial efforts to directly initiate development of a production code of the desired scope. As a result, a limited-scope prototype code development effort was undertaken with goals that were more modest yet essential to the overall success of the xLPR project. The Pilot Study had a defined timeframe for completion, and the objectives were explicit and limited. The scope was narrowly focused on assessing the probability of rupture of dissimilar metal pressurizer surge nozzle welds (specifically, those welds assessed in a prior program resulting in considerable relevant, publicly available technical data). A general description of key elements of this prototype development effort follows.

2.1 Project Goals

With three primary goals, the xLPR Pilot Study was undertaken as a step toward initial development of a production xLPR code:

(1) Exercise the proposed program organizational structure and assess its appropriateness to guide the development of a complex probabilistic computer code.

(2) Determine the computational framework structure that is most appropriate for the production xLPR code.

(3) Demonstrate the basic analytical and computational feasibility of the proposed probabilistic evaluation of pipe rupture.

Under the auspices of the addendum to the memorandum of understanding between the NRC and EPRI [1], a project organization with balanced representation from the two sponsoring constituencies was put in place to manage the effort and provide project direction. However, line authority over direction of the technical work of the xLPR project remained vested in the project managers of NRC RES and EPRI MRP through their oversight of xLPR contracts. Managing interfaces and communications, directing work effectively, developing an effective decision-making process, and obtaining active engagement in the project by individuals with the diverse range of technical expertise necessary to competently and thoughtfully execute the work of xLPR all present organizational challenges. Successfully navigating these issues in a typical corporate organizational structure with clear lines of authority can be challenging. However, overcoming these issues in a diverse, informal, matrixed organization must rely on "softer" team attributes including compatibility of key participants, force of individual personalities, and collective interest in the success of the enterprise. Therefore, as stated in the first goal, it has been essential to exercise the technical and managerial organizational structure initially set up for the xLPR project to determine whether it is sufficiently robust to successfully manage this multiyear, multi-organization, technically complex methodology and software development effort.

Determining the most appropriate computational framework for the full xLPR code represents a fundamental and particularly challenging decision for the project and consequently was the second key goal for the Pilot Study. An initial constraint placed on the code development process was to maintain substantial flexibility to allow the production versions of the code to evolve with relative ease as experience with and knowledge of the fundamental physical phenomena evolve. Retaining the ability to readily incorporate newly developed or refined models for materials degradation compels the selection of a modular framework over more traditional hard-coded software approaches. However, beyond this fundamental selection, many important attributes distinguish modular framework platform approaches and demand additional thoughtful evaluation. Development of an open source, unlicensed platform for the project versus code implementation within a commercially licensed platform was identified early as the second-tier key distinguishing attribute for platform selection. The complexity and implications of this selection decision for the entire project suggested that application experience in each of these framework development environments would provide the most relevant and useful input to the evaluation process. Thus, the Pilot Study involved development of two functional computational tools built on comparable platforms, one licensed and the other developed entirely with open source software elements.

Demonstrating the basic analytical and computational feasibility of this proposed probabilistic evaluation of pipe rupture was the ultimate goal of the Pilot Study, since failing this test makes organizational viability and platform selection superfluous. Meeting this third goal necessarily involves development of a functional framework code that is then exercised on a defined sample problem. Within this broader goal, major elements include the following:

- Extremely low probabilities of rupture can be calculated.

- A methodology for calculating the distribution of this probability can be developed.

- The sources of uncertainty in this distribution can be identified and quantified and are appropriately characterized and propagated throughout the analysis.

- The capability for a limited sensitivity study of key analytical drivers is provided to aid identification of technical areas where the current state of knowledge might be inadequate to support further xLPR development.

- The appropriate documentation for the adoption of a model, input, probability density function, and other aspects exists and is controlled.

The project team has gained much experience through the Pilot Study, as summarized here and documented in the supporting references, which will now benefit the future development of the xLPR production code.

2.2 Pilot Study Organization

The organizational structure put in place to conduct the Pilot Study reflects the reality of two funding entities, perceived challenges with co-mingled funds, desire for relatively balanced project oversight by the dual sponsors, and recognition that project success depends on competent treatment of the technical issues. The resulting structure consisted of four technical task groups reporting to a management committee. The technical task groups and highlights of their areas of responsibility are the following:

- *Computational Task Group*: integration of the computational elements (models) into a robust, fully developed, tested, and verified computational tool

- *Models Task Group*: selection, documentation, and coding of the mathematical model building blocks

- *Inputs Task Group*: identification, collection, and presentation of the data required for both the models and the sample problem as input tables and distributions

- *Acceptance Criteria Task Group*: formulation of probabilistic acceptance criteria for assessing the implications of the results and overseeing the evaluation of the dual Pilot Study framework codes

Co-chairs representing both the NRC and industry constituencies led each of these technical task groups. Task group membership consisted of staff or contractors with relevant expertise drawn from, and separately funded by, either the NRC RES or EPRI MRP.

The two co-chairs of the four technical task groups, plus a management representative and an at large member from each of the constituencies, composed the management committee, designated the Project Integration Board (PIB), for a total of 12 members. This group was invested with general oversight of the project, specifically including the project scope and technical direction. However, in a traditional line management approach, day-to-day decisions and all financial and contractual actions necessarily remained the responsibility of the several NRC and EPRI project managers (and PIB members) with authority over the various xLPR contracts.

2.3 Code Development Process

The Computational Task Group led the Pilot Study code development process. Models of the relevant physical phenomena were developed as independent modules, allowing project collaborators to work in any programming language. These modules were then directly coupled to the xLPR framework using dynamic link libraries (DLLs) by wrapping the original module source code in a simple standard dynamic link library shell.

Because of the limited time allowed for the Pilot Study, it was imperative that the framework development process not be delayed by a potentially protracted model selection process, so members of the Computational Task Group exercised their best judgment to select an initial set of readily available models for all necessary computational elements of the code. These models, designated as the "Alpha" models, typically were already available as existing subroutines in various legacy codes and could be easily adapted to the needs of xLPR with minimal native programming. This allowed work to begin on building the framework without having to guess the functional and interface attributes of the computational elements being linked.

The Models Task Group had the assignment of systematically and thoughtfully reviewing models for each of the computational elements to either ratify the initial Alpha model selection or identify a more appropriate option that was still consistent with the narrow focus and time constraints of the Pilot Study. In most cases, the Alpha selections were accepted as reasonable choices for the purposes of the Pilot Study. However, the final "Beta" set of models used for Version 1.0 of the codes did include a few changes and additions (see Section 3.2). This review process also identified significant issues and decisions that must be addressed in subsequent phases of the project.

2.4 Platform Selection

To fulfill the xLPR Pilot Study computational platform evaluation objective, both commercial and open source framework software was considered.

2.4.1 Commercial Software

Sandia National Laboratories (SNL) conducted an initial review of available modular computational platforms against the general requirements and expected needs of the xLPR project [2]. The commercial software recommended as most suitable for xLPR is GoldSim Pro, a dynamic, probabilistic simulation package with extensive capabilities relevant to the technical, developmental, ease of end-product use, and project controls scope of xLPR.

Developed by GoldSim Technology Group, LLC (GTG), GoldSim Pro is a general purpose, dynamic, probabilistic simulation software, which includes both a model developer's version and the simulation software. GoldSim Pro is compatible with the free downloadable version of the GoldSim Player software, which allows the user to view and navigate through the model logic, run an existing GoldSim model, modify input values and model options, and display the results without having to purchase GoldSim Pro. The xLPR framework model player file was created such that key inputs to the model can be modified before running the code.

The modular-based GoldSim framework model for the Version 1.0 xLPR model serves as the integrating shell that manages input variables (e.g., material properties) and model output (e.g., results), as well as the flow of information that includes the system-level model logic. It controls the order in which the modules are called and the passing of variables into and out of modules. The GoldSim framework for xLPR was constructed with an option to use standard Microsoft Excel spreadsheets to define the inputs as well as dynamically pass simulation results to Microsoft Excel for advanced post processing.

The framework uses the GoldSim software libraries of probability distribution functions and the capability to correlate variables and perform multiple-realization stochastic analyses in a Monte Carlo approach. The framework benefits from the GoldSim software's ability to store simulation data from large numbers of realizations and generate statistics on global probability distributions. GoldSim permits each run to be saved in a single action, including all input data and results from Monte Carlo analysis. Finally, the GoldSim framework has built in graphical user interface (GUI) functions that allow the developer to quickly assemble specific model runs and to create interactive player files for end-users.

The GoldSim software provides a visual and hierarchical modeling environment, in which the xLPR framework model was constructed by adding "elements" (model objects) native to the software that represent data, equations, module interface, processes, or events and linking them into graphical representations that resemble influence diagrams. These visual representations and hierarchical structures help users to build very large, complex models that can still be explained to interested stakeholders (e.g., government regulators, elected officials, and the public).

In addition, the GoldSim framework for xLPR includes the software's ability to track changes that have been made to a model file. This feature (referred to as "versioning") allows the differences between the current version and a previous version of a model file to be quickly determined. The version history is an integral part of the model file, providing an easy-to-access history of all of the changes that have occurred over the life of the model. Providing this configuration management (CM) capability is particularly useful for coordinating model changes when multiple people have the ability to access and modify the same model file and as a QA/quality control feature allowing for verification and documentation of where and when a model file has been changed.

2.4.2 Open Source Software

Oak Ridge National Laboratory (ORNL) had the task of developing an open source platform capable of supporting the needs of xLPR, as demonstrated through use in the Pilot Study. The structure and capabilities of this purpose-built simulation platform, named Structural Integrity Assessment Modular—Probabilistic Fracture Mechanics (SIAM-PFM), while not exclusively developed for xLPR, were substantially influenced by the needs of xLPR.

The SIAM-PFM (SIAM for short) framework is a problem-solving environment. SIAM-PFM is an object-oriented open source framework, within which a wide range of nuclear power plant safety issues can be addressed systematically and consistently by using modern principles of probabilistic risk assessment. This platform is readily extensible to different problem classes.

Every SIAM-PFM component is written using the Python programming language and Python frameworks. Components are easily installed and uninstalled on Windows operating systems, and all components are potentially portable to other operating systems.

In the SIAM-PFM problem-solving environment, all components use the same working principle: workspaces that contain projects (SIAM projects) or directories, in which all inputs and outputs of a given test case are saved. Users can navigate through the different projects in the project-explorer panel and create projects that represent test cases. Convenient plots of the outputs are also provided to visualize data, and users can also extract the raw data in text files to create custom plots.

The implementation of xLPR within SIAM (SIAM-xLPR) presents a series of tabs to define the case conditions. In these tabs, predefined input values have been set by default to the probabilistic base case. The SIAM-xLPR main GUI framework has seven tabs. Input data can be entered on the first six tabs, in any order. Default input values are provided and can be modified as needed. The seventh tab displays the SIAM-xLPR "Execute Utility" window, which presents a command line view of program executions.

Each of these two selected platforms is described in greater detail in [5] and fully described in their respective final reports [2, 3].

2.5 Configuration Management

Within the context of the Pilot Study code development process, the Computational Task Group imposed a moderately rigorous set of CM controls on essentially all project activities. However, these controls were not strictly linked to a formal QA program. CM controls and software QA program elements can be integrated into, and overlaid upon, a software development project to achieve specific goals. These goals may range from simple compliance with externally imposed requirements to fulfilling valued internal project needs for well-defined design bases, structured communications, thorough development and validation documentation, and effective internal project controls. With a clear understanding of project expectations from a QA program, a set of

CM/QA implementation elements can be developed consistent with those specific needs and requirements.

The xLPR Pilot Study is considered only a first step, and its results are not intended to be used in any formal manner external to the xLPR project itself. However, while no formal QA program requirements were imposed, implementation of a relatively rigorous CM program was judged to be of significant value to the project for several reasons. The process rigor imposed by even a modest CM program benefits the overall project in terms of controlled, consistent, and traceable documentation of the project elements, added confidence derived from peer-checking of inputs, and improved project coordination realized through version control, particularly in a project with many participants such as xLPR. However, in the particular case of the Pilot Study, there is a very real possibility that elements from this developmental stepping stone effort will also become building blocks in the production code. Since it is expected that xLPR will be used to support licensing actions by both the NRC and licensees, software QA fully compliant with relevant NRC and industry requirements must be imposed during the production code development process. Therefore, imposing a reasonably comprehensive suite of CM controls on the Pilot Study, although still less comprehensive than a full software QA program might demand, establishes the mindset of working in a CM environment, enforces solid project documentation, and will substantially ease the incorporation of Pilot Study elements into the QA environment that will be defined for production code development. Furthermore, this Pilot Study CM experience provided practical, directly relevant input to the determination of the precise QA/CM program details that are going to be imposed on the production code.

CM is the process of identifying and defining the important configuration items in the system, controlling the release and change of these items, reporting their status over the course of the project, and verifying their completeness and correctness. The CM plan implemented throughout the Pilot Study applied a systematic approach to both the developed software and models to ensure that the fundamentals of software CM were met, including the following:

- access control
- version control
- verification (e.g., checking)
- traceability (e.g., documentation)

The CM process was implemented from a series of guidance documents that outline the specific steps for each of four key components of the xLPR pilot program:

(1) module development

(2) framework development

(3) model parameters and inputs for the Pilot Study test case

(4) xLPR model production runs and uncertainty and sensitivity analyses for the Pilot Study test case

3
CODE STRUCTURE OVERVIEW—XLPR VERSION 1.0

The complexity associated with evaluating the likelihood of piping rupture derives from the uncertainty with which each of the input parameters is known and how well the relevant physical phenomena are understood and modeled. Fundamentally, a probabilistic Monte Carlo type analysis combines and propagates these uncertainties by sampling from the statistical distribution for each input (including inputs describing model uncertainty) over its defined range and executing the prescribed analysis, then repeating this process many times. The results form a probability distribution of their own that characterizes the performance of the system. These results then may also be evaluated to draw insights into the apparent correlations between specific input parameters (or sets of parameters) and the expected outcome. Such insights may be helpful in better understanding system behavior, as well as informing critical decisions regarding system design, operation, maintenance, inspection, and regulation.

The flow of the xLPR Version 1.0 code is centered on a time-based history of events where PWSCC flaws initiate and grow until failure. This simulation structure was developed within the Computational and Models Task Groups, which coded, compiled, and, using the project-defined CM process, verified the set of computational modules needed for this purpose. The individual modules address loads (including weld residual stress), crack initiation, crack growth, crack coalescence, crack stability, crack opening displacement, leakage, inspection, and mitigation. Both the commercial and open source applications used these self-contained modules which were linked to create xLPR framework codes that control the time flow of the analyses and properly account for and propagate the problem uncertainties. While the two framework codes do use the same analytical model modules and implement essentially the same time loop structure, platform functionalities and programming preferences did result in generally minor differences between the two frameworks.

3.1   Uncertainty Characterization

Within the Pilot Study, uncertainty was characterized as either aleatory or epistemic according to the following definitions:

Aleatory Uncertainty: This type of uncertainty arises because of natural, unpredictable variation in the performance of the system under study. The knowledge of experts cannot be expected to reduce aleatory uncertainty, although their knowledge may be useful in quantifying the uncertainty. Thus, this type of uncertainty is sometimes referred to as irreducible.

Epistemic Uncertainty: This type of uncertainty results from a lack of knowledge about the behavior of the system that is conceptually resolvable. Epistemic uncertainty can, in principle, be eliminated with sufficient study, and expert judgments may be useful in its reduction. Epistemic, or internal, uncertainty reflects the possibility of errors in our general knowledge.

In the context of xLPR, the *probability of pipe rupture* is controlled by randomness in the behavior or properties of the piping system (aleatory uncertainty). The *uncertainty in the probability of rupture* is driven by the lack of knowledge with respect to quantities used in the calculation of probabilities of rupture that are assumed to have fixed but imprecisely known values. Attempting to distinguish between these uncertainty types is expected to be important to xLPR because the overall degree of uncertainty inherent in the pipe rupture problem is considered to be relatively high. Gaining some sense of the extent to which that uncertainty may be inherently random or might be further reduced may better inform current regulatory decisions, as well as lead to better use of future research dollars.

3.1.1 Uncertainty Propagation and Sampling

The basic architecture of the xLPR framework is a set of nested loops used to separate the epistemic and aleatory uncertainties. For each cycle through the outer loop, a single sample for each of the epistemic parameters is selected and held constant while the action passes to the inner loop. Within the inner loop, the aleatory parameters are sampled each iteration, the analysis involving all input parameters is executed, and then the process repeats for the desired number of inner loop iterations (NA). The outer loop then enters a new cycle, a new set of epistemic parameters is selected, the inner loop iterates NA times, and the process repeats for the total number of epistemic realizations (NE). Thus, each epistemic outer loop cycle has NA number of possible outcomes and (NE x NA) represents the total number of possible outcomes generated in the model simulation. Each epistemic realization output then is a distribution of the results based on the aleatory samples contained within it. Further insights can also be gained by selectively switching key parameters from aleatory to epistemic (or vice versa), rerunning the analysis, and evaluating any resulting differences in the results.

In most circumstances, it is suspected that the calculated probability of rupture for the subject piping systems would be extremely low since the prevailing design requirements and constraints, as well as operating experience, suggest that the actual probability of rupture is also extremely low. Random sampling of the uncertain input parameters will therefore result in very few realizations with rupture predicted. Consequently, the calculated probability of rupture will be determined by those relatively few runs and thus will be poorly estimated. Furthermore, any correlation with relevant input parameters will be poorly characterized. Therefore, several random variable sampling techniques were incorporated into the Pilot Study codes to evaluate their efficacy in more efficiently sampling the key input parameters to improve the resolution of rupture probabilities within reasonable run times. In one case, the method adaptively focuses sampling in statistically valid ways within input parameter ranges of greater relevance to the low–probability events of interest. These advanced sampling approaches are more fully described in [5].

3.1.2 Time Loop

As described above, the uncertainty propagation structure for the Version 1.0 xLPR code consists of an inner aleatory loop and an outer epistemic loop (Figure 3-1). Each random variable is assigned to and sampled within one of these two loops. After all parameters are sampled, the input conditions are constructed to form a timeline for that particular realization, and execution enters the time loop depicted in Figure 3-2.

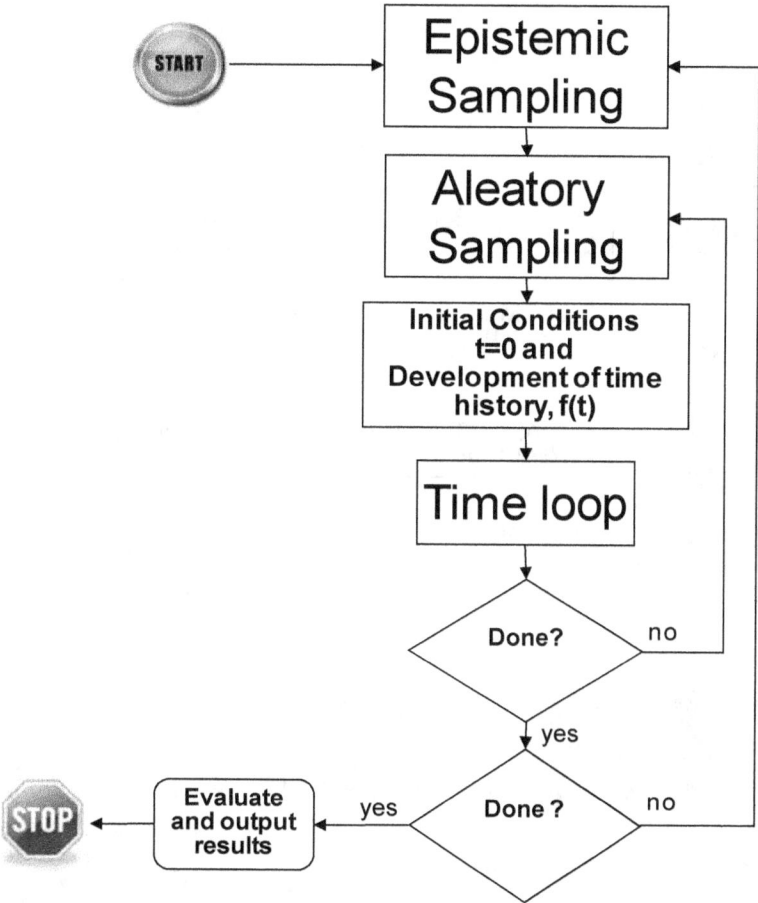

Figure 3-1 Schematic main loop (outer) flowchart for Version 1.0 xLPR code

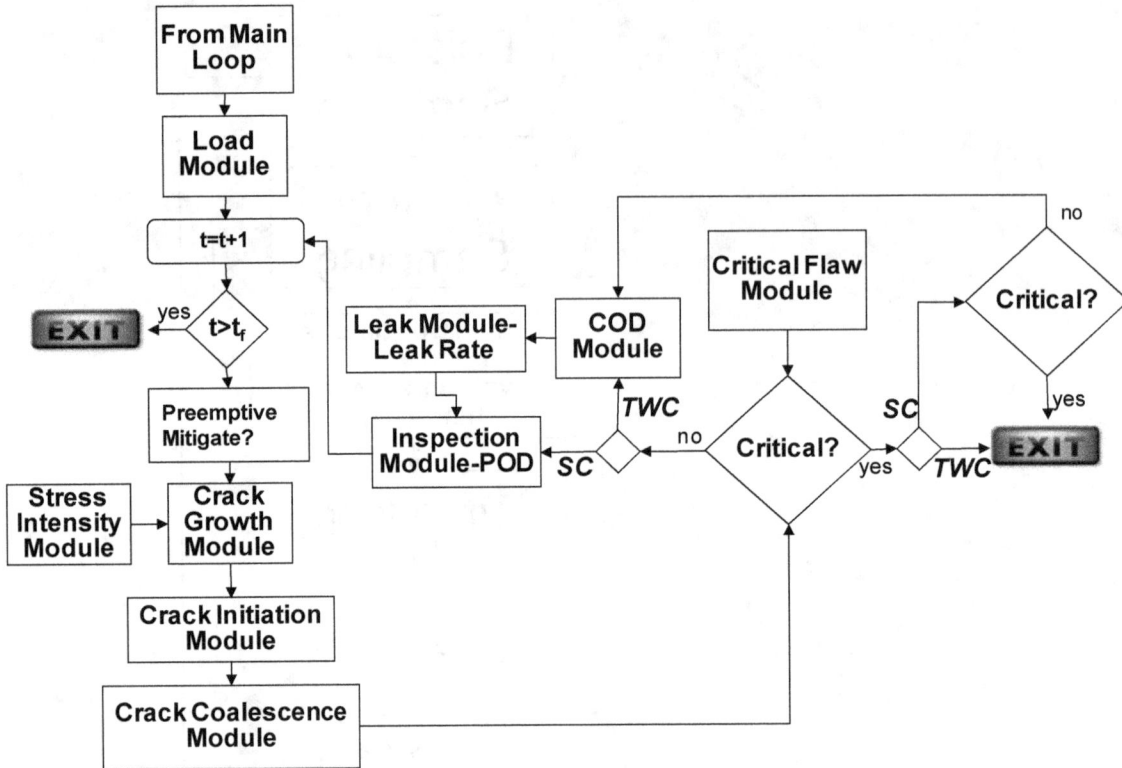

Figure 3-2 Schematic time loop (inner) flowchart for Version 1.0 xLPR code

3.1.3 Outputs and Post Processing

To maximize efficiency and extract as much information from a complete run as possible, only rupture will result in exiting the time loop before reaching the predefined time limit. This allows data on leakage and inspection detection probabilities in particular to be recorded over the entire time period so that various inspection criteria and detection thresholds can be evaluated after the run is complete by post processing the results. Details of the various outputs and the post-processing routine can be found in [5].

3.2 Module High-Level Overview and Limitations

3.2.1 Pilot Study Model Selection

The individual modules, referenced above in Figure 3-2 as to their basic function, are discussed more fully in [5] and described in detail in [4]. The mathematical models of the underlying physical processes of interest within each module were selected for the Pilot Study generally as a practical balance of their veracity, availability, and compatibility with xLPR objectives and companion models.

As the initial step in the selection process, the Models Task Group identified the available modeling approaches in each of the required analytical areas. A robust set of screening criteria [4] was developed to assess their usefulness and applicability for the intent of the Pilot Study in particular and xLPR in general. The veracity with which they represent the underlying physical phenomena, their benchmarking performance against available physical data, and their acceptance within the scientific community were all identified as important model selection criteria. The availability of the model for use in this project and whether it could be obtained as a coded computer subroutine, or could easily be coded, were of particular interest for the Pilot Study because of its accelerated schedule. Candidate models ranged from simple to complex, and in each area, their compatibility with project goals, framework structure, and computational constraints had to be considered. Exceedingly complex models might represent the underlying physical process more closely but be impractical to implement because of a lack of necessary input data or excessive computational demands, or because a simpler model was sufficient considering the parameter's importance within the overall Pilot Study analysis.

However, the urgency of beginning framework development within the accelerated Pilot Study schedule and the inherently complex interaction between the models and framework necessitated early selection of a set of placeholder ("Alpha") models by the Computational Task Group. While these early selections prejudiced the final models selection process to a degree, they were competently made and thus the Models Task Group ultimately ratified many of them.

In most cases, the Models Task Group followed a consensus process in model selection that involved most of the formal selection criteria as points of discussion, but the structured model evaluation and selection process as initially conceived was not completed for the Pilot Study. For those models for which a widely accepted, readily available, already coded model had been selected within the "Alpha" model set, less discussion or "selection" was warranted. For other models, for which no well-vetted and accepted models existed, the selection process was more difficult. Nevertheless, the final set of selected models was fully documented in accordance with the requirements of the CM program and is described in detail in [4]. It is expected that a more formal and deliberate model selection and development process will be exercised for Version 2.0 of the xLPR code.

In the Pilot Study, only one model was chosen for each calculation, except for defining crack initiation, for which three models were chosen and coded. Discussions of Version 2.0 of xLPR code development have considered making available several validated models representing more of the computational elements, thus leaving the selection to the user for comparative or sensitivity studies.

3.2.2 Pilot Study Model Limitations

Because of the limited scope of the Pilot Study problem, some limitations were deemed acceptable in model selection that will, in most cases, not be acceptable within the production code and thus will demand further consideration in subsequent phases of the xLPR project. The following list briefly summarizes the more noteworthy Pilot Study model limitations:

- Manufacturing and welding defect distributions are not included or considered.

- Fatigue flaw initiation or growth is not considered (no transient load definition).

- The model for piping loads is simplistic and thus may not be sufficiently representative.

- Only circumferential surface-breaking cracks are considered (no axial cracks).

- Only idealized crack behavior is considered.

- The crack transition model from part-through to through-wall is simplistic.

- The surface crack stability model to predict rupture is simplistic.

- A simplistic weld residual stress model is used, which is a third-order approximation and assumed constant around the circumference.

- Some important leak rate parameters were "hard-coded" instead of using inputs or dependent variables.

- A simplistic inspection model is currently implemented:

  - Sizing uncertainty must be addressed for repair or replace decisions.

  - A repair or remediation decision process needs to be developed.

  - Treatment of post-repair (new fabrication-induced) crack distribution is needed.

- A simplistic mitigation model is currently implemented (only "generic" stress modification); mitigation methods to be addressed include the following:

  - weld overlay (both full structural and optimized)

  - mechanical stress improvement

  - weld inlay/onlay

  - inner diameter surface treatment (including peening)

  - inspection-based/material replacement mitigation (e.g., Alloy 52/152 welds)

  - chemical mitigation should also be incorporated

Section 7 of this report presents a more detailed summary of issues and development opportunities identified during the Pilot Study that are relevant to future production versions of xLPR. This summary is based on detailed discussions in [2, 3, 4, 5].

The summary of issues identified above represents a daunting set of technical challenges that collectively may overwhelm the ability of the xLPR team to adequately address them simultaneously in the first production version of the code. Therefore, each technical issue identified within the modeling and framework development area of the Pilot Study will be carefully evaluated to determine both how critical its resolution is to the success of xLPR Version 2.0 and how practical it will be to resolve that issue within the constraints of the overall project schedule and available funding. In cases where substantial progress toward resolution may not be practical given these limitations, a lesser degree of progress may be considered in a continuing effort to improve the code. A detailed Version 2.0 program plan will be developed early in the process to document all decisions regarding scope definition and to guide future code development.

4
ORGANIZATION EVALUATION

The organizational structure employed for the Pilot Study, as described in Section 2, was established to impose a degree of order on the project activities yet still allow a free flow of ideas and involvement of a wide range of technical experts. The organization consisted of four technical task groups reporting to a management committee. Each group's work was directed by "co-chairmen" drawn respectively from the two funding constituencies, reflective of the overall cooperative structure of the project. The many technical subject matter experts who participated in the Pilot Study were similarly drawn from both the NRC and industry sources, thereby resulting in greater diversity of input and experience than would otherwise have been possible. However, by its very nature, this organization lacked a clear line of authority and was instead, in effect, structured around parallel lines of authority. Particularly in the earlier stages of a newly formed cooperative venture focused on a complex technical task such as the xLPR project, directly implementing a more formal organization can be challenging. The scope and details of the project may lack sufficient definition to contribute to organizational development, and a range of technical, managerial, and financial control issues must also be addressed. Therefore, the Pilot Study became an incubator for both the organizational and technical aspects of the project. By affording the participants an opportunity to gather, develop into a functioning team, and share in a common goal, the Pilot Study has allowed essential space for the project to evolve to a point that the more challenging organizational issues can now be addressed and resolved.

Over the course of the Pilot Study, the role of each group evolved and became better defined, as did the interactions between the groups. As initially conceived, the PIB was intended to provide general oversight and direction to the project and the technical task groups, arbitrate the competing roles and perspectives of the groups as project decisions were needed, and ensure effective communications within the organization. However, a loosely organized committee of 12 members could not effectively handle day-to-day decisions. Furthermore, consistent with traditional line management roles in the two funding organizations, all financial and contractual actions obviously remained the responsibility of the several NRC and EPRI project managers (who were also PIB members) with authority over the various xLPR contracts.

Also, given the well-defined, narrow scope of the Pilot Study and its relatively short duration, the "active direction" role of the PIB faded somewhat very early in the project, becoming secondary to direct interaction and practical decision making between the most active task groups. Development of the xLPR framework and integration of the code modules into fully functioning computational tools effectively dictated the overall project timeline. The Pilot Study analytical goal was focused more on overall computational feasibility rather than demonstrable accuracy so selection of best available, simplified models and inputs was judged sufficient. Consequently, the Computational Task Group began to drive the project consistent with the needs of framework development. This early evolution in organizational roles improved overall project efficiency, but had the effect of sidelining the PIB throughout much of the Pilot Study.

The Models Task Group had primary responsibility for selection of an appropriate set of mathematical models for the relevant physical phenomena included within the overall fracture mechanics-based computational methodology. The group developed a thoughtful set of evaluation criteria considering such factors as scientific veracity, compatibility with the overall methodology, availability for immediate use, and suitability for the problem at hand. However, the model evaluation, selection, and documentation process required time for the necessary organizational dynamics to take root and then to exercise the process. The initial framework development was a parallel activity, but could not proceed without at least a placeholder set of models for the essential code elements. Consequently, the Computational Task Group quickly selected readily available models for their immediate use and asked the Models Task Group to either ratify those selections for Pilot Study use or identify and provide fully coded replacements.

Each model represented a somewhat unique field of expertise, so the members split into modeling subgroups which met on an as-needed basis as the individual models were being evaluated or developed. However, these individual teams often had no direct interaction with the framework developers regarding model incorporation into the framework, which, in some cases, led to confusion. Fostering efficient communication between and within the task groups thus became essential to project success and was particularly difficult within the subgroup structure of the Models Task Group.

The Inputs Task Group, with responsibility for the many inputs required by the xLPR Version 1.0 code, had a particular challenge in defining appropriate distributions for numerous inputs that, in many cases, are not routinely described in that manner. Within the limited context of the Pilot Study, the basic input data were largely contained within a publicly available report [6]. Furthermore, for the purposes of the Pilot Study, it was acceptable to make assumptions regarding the associated distributions and even to use engineering judgment to select replacement data, when necessary, to stand in for missing data. However, while this resulted in the group being less active over the course of the Pilot Study, those opportunities to make assumptions and easily access the necessary data will not carry over to the production versions of the xLPR code. Acquisition of the necessary data from a broader range of weld joints, translating such data into appropriate distributions, and classifying the associated uncertainties in a defensible and consistent manner are among the added challenges faced by the Inputs Task Group for xLPR Version 2.0.

The Acceptance Criteria Task Group, initially intended to define appropriate and reasonable standards for the evaluation of xLPR results, was refocused for the Pilot Study on the task of evaluating the two comparative frameworks. Similar to the Inputs Task Group, this group's original function is significant within the overall xLPR project, but within the limited context of the Pilot Study was of less importance and thus not exercised to a meaningful degree.

Thus, the bulk of the Pilot Study effort was concentrated within the Computational and Models Task Groups and the framework development teams at SNL and ORNL. While this was, in the end, appropriate and sufficient to meet two key goals of the Pilot Study, it did not directly address the third goal of fully exercising the organizational structure. However, the very fact that the organizational roles and responsibilities evolved as they did addressed many important aspects of this goal.

While the Pilot Study organization proved to be sufficiently flexible to support and accommodate the evolving needs of the project, it is not clear that it can adequately meet the

5
PILOT STUDY PROBLEM RESULTS

5.1   Pilot Study Problem Statement

To meet the goal of demonstrating the basic analytical and computational feasibility of assessing the probability of pipe rupture as conceived within the xLPR project, a problem statement was developed for the Pilot Study. This problem statement consisted of two deterministic analyses and a probabilistic base case with a series of sensitivity analyses to demonstrate the features of the Version 1.0 code in the context of the pressurizer surge nozzle. The problem statement is presented in more detail in [5]. Each of the two functioning codes was exercised with this problem statement to address the feasibility goal and also to directly compare the usability and results from the two frameworks. The results presented here are not intended for formal use outside the xLPR project, and the absolute values of the probabilities have not been validated. However, relative results from comparable sensitivity runs can be quite informative for project planning purposes.

To verify that the codes are performing the deterministic calculations correctly, two separate deterministic analyses were defined with the following attributes and otherwise identical inputs:

> Deterministic Analysis 1: Single crack at time = 0 years, with no mitigation. The location of the crack is at the top of the weld ($\Phi = 0$ radians).

> Deterministic Analysis 2: Three cracks at time = 0 years, with no mitigation. This analysis is an extension of the first deterministic analysis with three cracks. The three cracks are the same size as for Deterministic Analysis 1. Their respective locations are $\Phi = 0$ radians, $\Phi = 0.6$ radians, and $\Phi = -1$ radian.

The controlled versions of both the xLPR Version 1.0 GoldSim and SIAM framework models developed for the xLPR Pilot Study were exercised with the prescribed deterministic test case inputs. As described in [5], these runs demonstrated that both codes produce approximately[1] the same output.

Table 5-1 summarizes the probabilistic base case and five sensitivity cases.

---

[1] The slight difference in results was attributed to the time-step implementation differences between the two codes.

Table 5-1: xLPR Version 1.0 Probabilistic Analyses

Analysis	Description
Probabilistic Base Case	Probabilistic base case analysis using Monte Carlo sampling—high weld residual stress with no inspection, mitigation, or leak detection.
Sensitivity Studies	
Stress Mitigation	Analyses evaluate different mitigation times and stress inputs. Three cases were run with mitigation at time 10, 20, and 40 years. Two welding residual stress inputs were considered.
Chemical Mitigation	Chemical effects of increasing the hydrogen concentration in the water on the crack growth module. Three hydrogen concentrations were evaluated. The base case hydrogen, 25 cc/kg-STP, was increased to 50 and 80 cc/kg-STP.
Crack Initiation	Considers the crack initiation model uncertainty. An alternative initiation model was exercised.
Safe End Evaluation	Considers stainless steel safe end weld which causes a through thickness bending stress that can reduce the tensile inner diameter stress. Alternative stresses representative of safe end addition effects were input.
Importance Sampling	Discrete probability distribution analysis with importance sampling using the safe end evaluation analysis was exercised.

In addition to the sensitivity analyses, the base case and the safe end evaluation case were post processed to take credit for leak detection and inspection. For each case, leak rate detection limits of 0.1, 1, 10, and 50 gallons per minute (gpm) were considered. The inspection schedules assumed were every 5, 10, 20, and 30 years.

The inclusion of these sensitivity cases was essential to extracting maximum value and insights from the Pilot Study effort. The analysis cases selected were intended to highlight the effect of key input parameters, uncertainty categorization, and sampling approaches on the outcome. These insights provide important guidance for planning the work scope and computational structure of the next phase of xLPR.

5.2 Summary of Results

The two development teams formally exercised the respective codes developed under the xLPR Pilot Study on the sample problem set previously described. The reports published separately by SNL [2] and ORNL [3] document the results of these runs in detail, and the Version 1.0 Final

Report [5] compares and summarizes the results. In addition, xLPR team members have exercised these codes on individual problems of interest and shared their observations and experiences informally within the Computational Task Group.

In the formal sample problem runs, the probabilistic base case, which contained high mean weld residual stress with no inspection, mitigation, or leak detection, produced relatively high and very comparable mean probabilities of rupture using both the GoldSim and SIAM frameworks. Further analyses suggest that the calculated results are stable and highly driven by the epistemic uncertainty. Parameter sensitivity studies conducted on the base case suggested that the weld residual stress and the crack initiation parameters (all characterized as epistemic) were controlling the uncertainty in the probability of rupture. The benefit of leak detection and inspection was less than expected, which was attributed to the high rate of crack growth and assumptions made regarding crack behavior.

Each of the problems in the Pilot Study was run with both the GoldSim and SIAM frameworks. This section presents comparisons of the results from the two codes. In certain cases, results from only one code are shown since the trends between the codes were similar. When this occurs, the results from the code that is used in any particular figure are identified by a model result designation (i.e., GSxLPRv1.0 (GoldSim) or SIAM_v1.0 (SIAM)). In addition, for the base case analyses, the total number of realizations was chosen to be 50,000 (1,000 epistemic each with 50 aleatory).

5.2.1 Base Case

Figure 5-1 shows the "probability of rupture" results for the base case. In this figure, the light grey vertical lines represent the probability of rupture for each epistemic realization (i.e., representing aleatory uncertainty). For each epistemic realization in this case, there were 50 aleatory realizations. The fact that each grey line is nearly vertical indicates that for any given epistemic realization, all 50 aleatory realizations as a group either ruptured (value = 1) or did not rupture (value = 0) (i.e., the aleatory uncertainty had no effect). The vertical lines are well distributed across the 60-year time period, which indicates that the epistemic uncertainty is controlling the behavior. In addition to the light grey lines, Figure 5-1 shows the mean and standard quantiles. For the base case, the mean value suggests that there is a 41 percent chance of pipe rupture in 60 years (720 months).

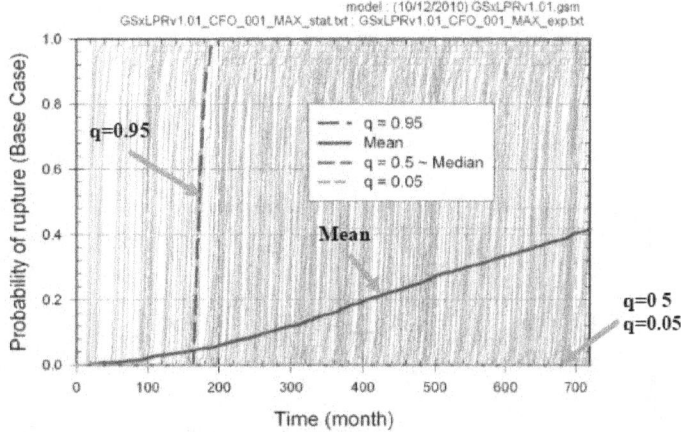

Figure 5-1 Probability of rupture for the base case

Switching the crack initiation parameters from epistemic to aleatory uncertainty greatly influences the behavior of the epistemic realizations, as shown in Figure 5-2. With reclassification of the crack initiation parameter to aleatory uncertainty, the probability of rupture varies with time (i.e., the grey lines are no longer vertical. This change leads to a smoother probability estimate over the 60-year timeframe. An interesting consequence shown by these assessments is that the quantile curves are now completely different. Because the time of crack occurrence was not fixed for each epistemic set, it is more likely to have at least one crack for each epistemic realization (but a smaller chance that *all* realizations within an epistemic set lead to rupture). While the estimate of the mean probability of rupture gives similar results, this is not the case for the quantile values. Their interpretation changes considerably from one assumption to the other. In Figure 5-2, for instance, a median of 0.32 at 60 years means that half of the epistemic realizations have at least a 32 percent chance of a rupture in the future, while in the base case, there was absolutely no chance of rupture for half of the epistemic realizations (the median in Figure 5-1 was zero at 60 years).

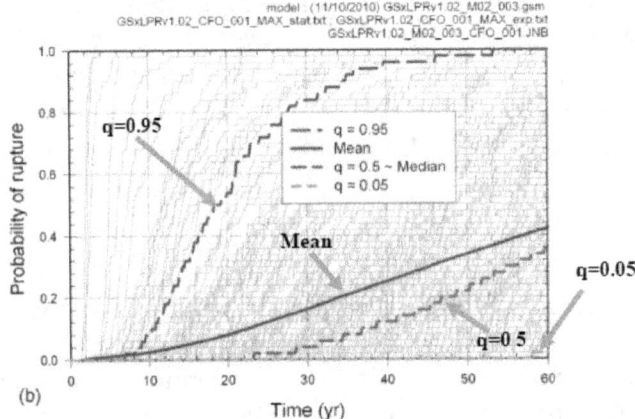

Figure 5-2 Probability of rupture for base case with crack initiation uncertainty changed from epistemic to aleatory

Figure 5-3 shows a comparison of the mean probability of rupture using the GoldSim and SIAM xLPR framework codes. The results show a good comparison between the codes, with the main difference resulting from slight differences[2] in the crack initiation probabilities.

---

[2] See [5] for more details.

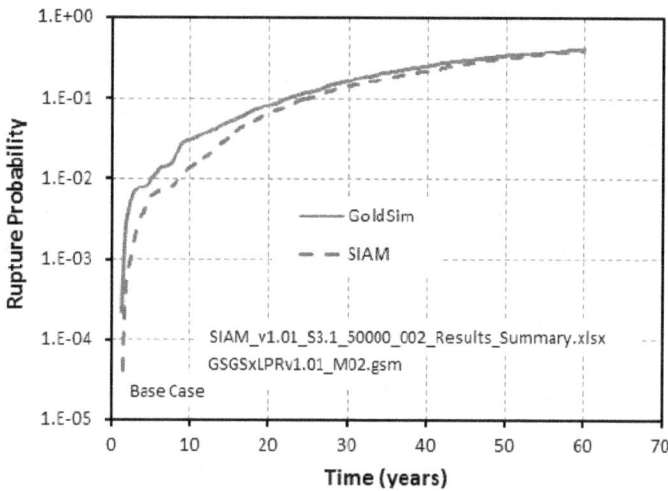

Figure 5-3 Comparison of GoldSim and SIAM framework for mean probability of rupture for the base case

Figure 5-4 shows the effect of the combination of leakage detection (on line) and periodic volumetric inspection (i.e., ultrasonic inspection) on the base case mean probability of rupture. In this case, a 2-year inspection interval and a leak detection limit of 1 gpm are assumed. These results illustrate a decrease of approximately 4 orders of magnitude in the mean probability of rupture when credit is taken for inspection and leak detection. This sensitivity case does not reflect the benefit of an assumed program of periodic bare-metal visual examinations for evidence of pressure boundary leakage performed during outages.

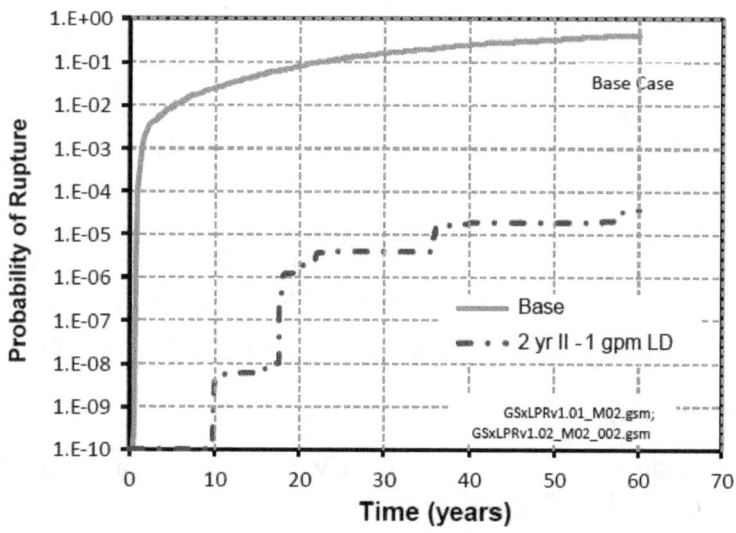

Figure 5-4 Effect of inspection interval (II) and leak detection (LD) on base case mean probability of rupture

5.2.2 Stress Mitigation

Figure 5-5 shows the effect of preemptive mitigation using a stress-based technique on the mean rupture probabilities. The only mechanical mitigation method incorporated in xLPR Version 1.0 is a preemptive stress-based mitigation. For this option, the user inputs a mitigated weld residual stress distribution and a time at which that mitigation is to occur. This modification of the stress profile affects both the crack initiation and growth models. For the example shown in Figure 5-5, the mitigation was applied at 10, 20, and 40 years. The results from these analyses demonstrate that, for the inputs considered, the application of the stress-based mitigation causes the mean probability of rupture to no longer increase with time. Since the data shown in Figure 5-5 represent the cumulative probability of rupture, a horizontal line represents no additional ruptures during that time period. However, a close investigation of the results from Figure 5-5 shows that before the mean probabilities cease to increase, they rise slightly above the non-mitigated mean rupture probabilities. This is because of the tensile zone in the mitigated weld residual stress distribution. For values between 0.5 and 0.9 for the ratio of crack depth to wall thickness, the mitigated weld residual stress becomes tensile. In those realizations where a crack is present that is at least 50 percent through wall at the time the mitigation occurs, the crack growth rate is increased. This effect causes the slight increase in the mean rupture probabilities before the mitigation effects become apparent. Finally, in practice, a volumetric examination for PWSCC flaws would typically be required at the time that a stress mitigation measure is applied in order to ensure that such deep flaws are not present in the subject component. However, such a test was not implemented within the Version 1.0 code.

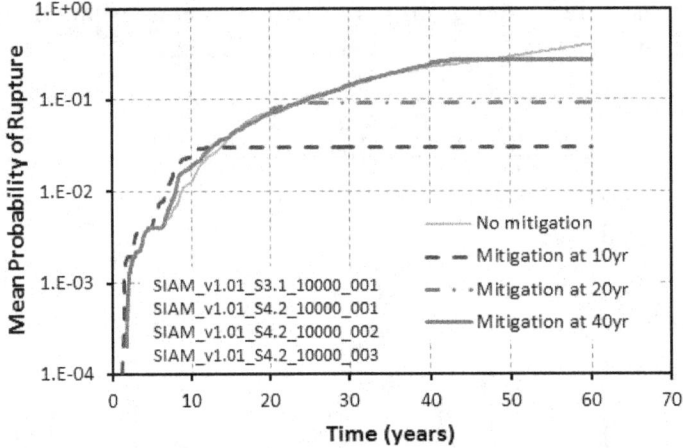

Figure 5-5 Effect of preemptive stress mitigation on the mean rupture probability

5.2.3 Chemical Mitigation

As described earlier, the effects of hydrogen on the PWSCC growth rate were implemented in the xLPR Version 1.0 code. The effects of hydrogen and zinc on the initiation of PWSCC were considered, but not implemented. Figure 5-6 shows the effects of increasing the hydrogen concentration on the mean rupture probabilities. The increase in hydrogen caused a decrease in the mean probability of rupture. This change is attributed solely to the change in the crack growth rate because of the increased hydrogen concentration. A large change in mean rupture probability occurred when the hydrogen content was increased from 25 cubic centimeters per kilogram (cc/kg) to 50 cc/kg. However, only a marginal increase in mean rupture probability occurred when the hydrogen concentration was increased from 50 cc/kg to 80 cc/kg. Overall, the decrease in mean rupture probability at 60 years is only about a factor of 2 when the hydrogen concentration increases from 25 cc/kg to 80 cc/kg.

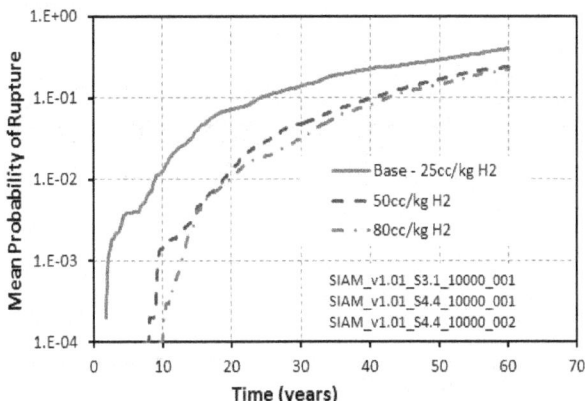

Figure 5-6 Effect of hydrogen on the mean probability of rupture[3]

5.2.4 Safe End Evaluations

Figure 5-7 compares the mean probability of rupture for the safe end sensitivity case to the base case. The only difference in the inputs for this case is the change in the weld residual stress. Inspections or leak detection are not credited in either case. The change in weld residual stress caused a decrease of 2 orders of magnitude in the mean probability of rupture.

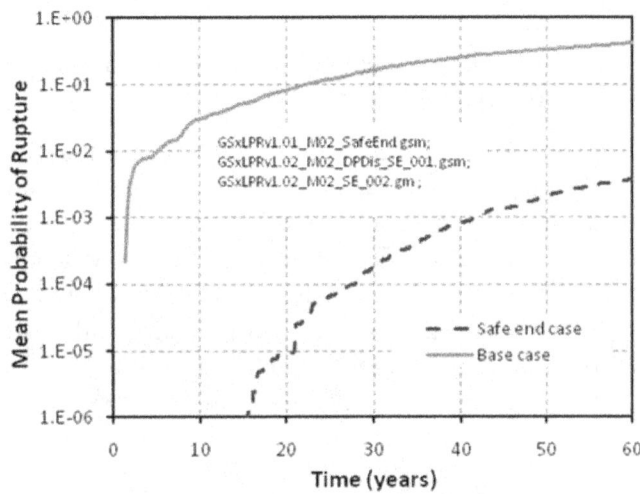

Figure 5-7 Mean probability of rupture for base and safe end case

Taking into account mitigation, a leak detection limit of 1 gpm, and a 10-year inspection interval reduces the mean rupture probability by 6 orders of magnitude at 60 years, as shown in Figure 5-8. For this example, a mitigation time of 20 years was chosen, and 10,000 realizations were used with importance sampling on both the weld residual stress and the crack initiation parameters, which were found to control the uncertainty in this problem. Figure 5-8 shows the

---

[3] Appendix D to [5] identifies an error in the hydrogen case runs that suggests the impact of H2 may be lower than this figure indicates.

results of the analyses. As shown in this figure, the mitigation at 20 years reduced the mean rupture probability at 60 years by 2 orders of magnitude, while the leak detection and inspections reduced the mean rupture probabilities by about 3 orders of magnitude. The combined effect caused a reduction of 6 orders of magnitude on the mean rupture probability at 60 years.

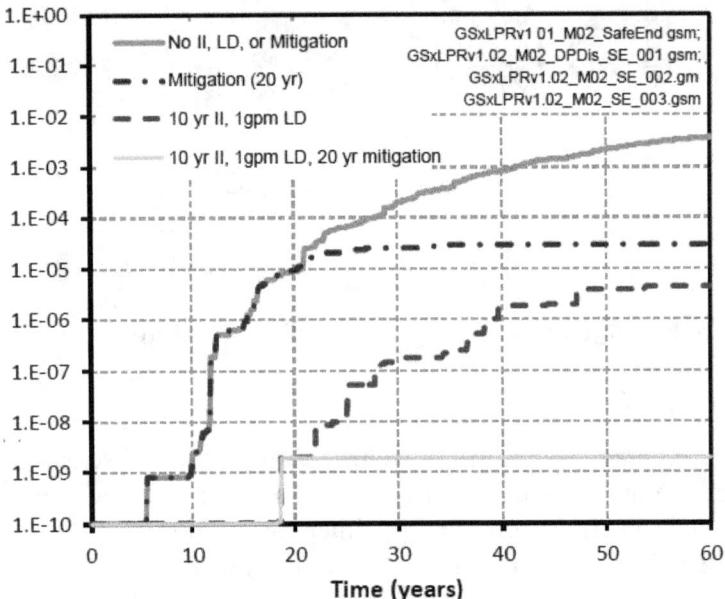

Figure 5-8 Mean probability of rupture for safe end sensitivity case with mitigation, leak detection (LD), and inspection interval (II)

The confidence for the mean value of the probability of rupture is calculated using the bootstrap method [5], which consists of sampling with replacement over the response generated by the original analyses. The central limit theorem states that when the mean and variance of the initial distribution are finite, the mean distribution should be asymptotically normal. Therefore, if the distribution of the mean values is normal, the sample size should be large enough to represent a stable solution. Figure 5-9 shows the confidence in the mean probability of rupture at 60 years for the safe end case with a 10-year inspection interval and a 1-gpm leak detection limit. The distribution in Figure 5-9 is nearly normal, which suggests a stable solution.

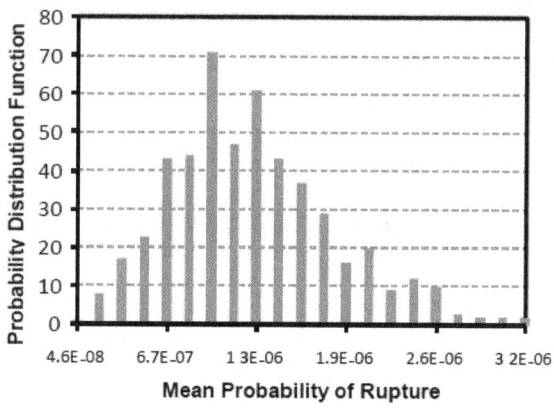

Figure 5-9 Confidence in the mean probability of rupture for safe end case with inspection and leak detection

Figure 5-10 shows the confidence for the safe end case at 60 years with mitigation at 20 years, a 10-year inspection interval, and a 1-gpm leak detection limit. Since only a few of the 10,000 realizations produced rupture with mitigation, inspection, and leak detection, the confidence in the mean value is low. As with the previous example, the confidence is estimated using the bootstrap method, which would produce a normal distribution of the mean if the analysis was stable. However, the results in Figure 5-10 suggest an exponential distribution and indicate a lack of stability in the analyses. Additional realizations would be needed to obtain a better estimate of the mean value for the probability of rupture. However, the distribution reported in Figure 5-10 spans approximately 1 order of magnitude, which is considered good based on the sample size (10^4) compared to the calculated probability of rupture (10^{-9}–10^{-8}). Based on the previous results, increasing the number of realizations leading to rupture by a factor of 2 or 3 may be enough to significantly increase the accuracy. Therefore, it may not be necessary to increase the sample size by an order of magnitude.

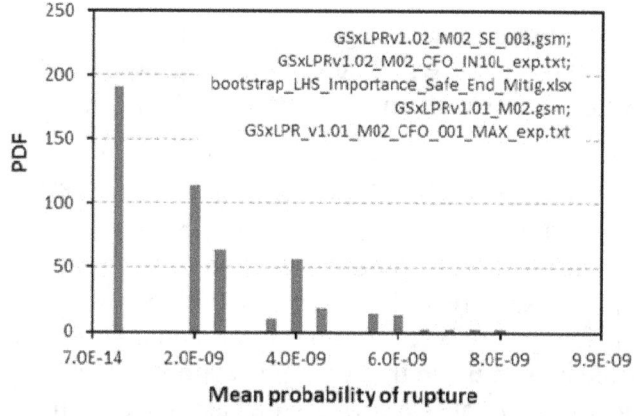

Figure 5-10 Confidence in the mean probability of rupture for safe end case with mitigation, inspection, and leak detection

5.2.5 Sample Problem Observations

The sensitivity studies conducted investigated the effects of stress mitigation, chemical mitigation, crack initiation model selection, and weld residual stress on the probability of rupture as compared to the base case. The conclusions from these sensitivity studies indicate the following:

- For stress mitigation, the effects are seen in both the crack initiation and growth models. The mitigation was applied at 10, 20, or 40 years. The results from these analyses demonstrate that the effective application of the assumed stress-based mitigation could cause the mean probability of rupture to no longer increase with operating time after its application.

- Two chemical mitigation cases changed the hydrogen content in the water from the base case value of 25 cc/kg to 50 cc/kg and 80 cc/kg respectively. As expected, these sensitivity cases indicated a benefit in reducing the mean probability of rupture because of the effect of hydrogen concentration on the PWSCC crack growth rate.

- Leak detection over the range of 1 to 10 gpm reduced the mean rupture probability at 60 years by a factor of 10.

- Inspection every 2 years reduced the 60-year mean rupture probability by a factor of 70, while inspection every 10 years reduced it by only a factor of 2.

- Changing the crack initiation model had very little impact on the overall mean rupture probability. This was expected since each of the models is empirically based and calibrated to the same service history.

- Changing the crack initiation parameters from epistemic to aleatory had a large effect on the probability of rupture distribution. While the mean value stays the same, when the uncertainty in initiation time is characterized as epistemic, there is a 0 percent chance of any future rupture 50 percent of the time. When the uncertainty is characterized as aleatory, there is at least a 32 percent chance of any future rupture 50 percent of the time.

- Changing the weld residual stress had a large impact on the mean rupture probability. Changing the residual stress from the geometry of a surge nozzle without a safe end weld to one with a safe end weld decreased the mean rupture probabilities at 60 years by 2 orders of magnitude.

The stability of the results was also evaluated to assess the efficacy of the sampling methods used. In many of the cases with low residual stress and/or inspection and leak detection, the number of realizations was insufficient to produce stable results. Therefore, importance sampling was needed to avoid excessively long run times driven by much higher rates of sampling. In all cases considered, the weld residual stress and an initiation parameter (B1) were importance sampled. The results indicate that with inspection and leak detection, mean probabilities of rupture down to 10^{-6} at 60 years can be calculated with reasonable confidence. However, when mitigation is added, mean rupture probabilities down to 10^{-9} at 60 years are calculated, but the confidence in the mean values is very poor. Additional realizations (predictions of rupture) are required to increase the confidence in these results.

Although these Pilot Study results provide interesting insights into possible trends regarding pipe rupture and key drivers, the overall pedigree of the two Version 1.0 codes does not support their use to draw broad conclusions outside the original scope and intent of the Pilot Study. While exercising these codes as part of both formal evaluations and informal familiarization, the project team identified and documented ease-of-use issues, opportunities for improvement, and a few errors for consideration, as appropriate, under subsequent phases of xLPR. However, the third goal of the xLPR Pilot Study, to "demonstrate basic analytical and computational feasibility for the proposed probabilistic evaluation of pipe rupture," was satisfied.

5.3   Sample Problem Conclusions

Exercising the two framework codes through the Pilot Study sample problem base and sensitivity cases gave the project team information about the overall xLPR concept feasibility goal, as well as many specific modeling, code structure, and system response topic areas. However, a few conclusions stand out as particularly significant:

- The results from the analyses support the feasibility of the computational approach to determining the rupture probabilities of piping systems.

- The importance of proper uncertainty characterization cannot be overstated.

- Initiation and weld residual stress can dominate the outcome of a problem driven by PWSCC; therefore, their modeling and treatment within xLPR must reflect the best available knowledge.

6
FRAMEWORK COMPARISON AND EVALUATION

The second Pilot Study objective, to determine the most appropriate computational framework, when combined with the stated goal of retaining modular flexibility in the code, largely predetermines use of a computational platform versus more traditional hard-coded, limited purpose approaches. Nevertheless, a range of computational platform options with various strengths and weaknesses could satisfy these requirements and thus needed to be evaluated. Among the many attributes to consider, open source (no license fees) versus commercially licensed software was judged to be a key selection attribute of sufficient importance to the project to warrant code development under both options for evaluation in the Pilot Study. A suitable open source platform code was already under development in a companion NRC-sponsored program at ORNL. Participation in the development of this code, known by the acronym SIAM-PFM (or just SIAM), provided an effective path to exercise and evaluate the open source platform option. A parallel effort was initiated with SNL building on its experience in other simulation projects using several commercially licensed framework codes. SNL staff screened likely candidate codes [2] and selected GTG's GoldSim as the best match of functional capability and user interface features to the needs of the xLPR project. GoldSim is a flexible platform for visualizing and dynamically simulating a wide range of systems and offers a free downloadable "player" application that allows GoldSim models to be viewed and run without requiring purchase of a license. However, GoldSim model development requires purchase of a software license. These two basic platform choices are each described in much greater detail in the Computational Group's supporting report [5] and in the individual reports prepared by the two framework code developers [2, 3].

6.1   Code Flexibility

The flexible structure of GoldSim and the inherent "flexibility" of SIAM as a software product still generally under development present a unique challenge for comparing and contrasting the two platforms. Fundamental limitations or important features that may distinguish the two platforms, but cannot be overcome in one or incorporated into the other with reasonable effort, may be rare. Therefore, an evaluation process was developed to attempt to identify more fundamental platform differences in usability and functionality, beyond just a simple direct comparison of the Pilot Study framework implementations in the two platforms. An independent report prepared by the Center for Nuclear Waste Regulatory Analyses (CNWRA) presents details of this process and its results [7]. The conclusions drawn both from this formal evaluation process and other xLPR project inputs are discussed in detail below.

6.2   Code Modularity

The concept of a truly modular approach to the code has been almost a mantra from the inception of xLPR, but the Pilot Study has exposed the practical realities and limitations associated with meeting this goal. True modularity offers the hope of allowing a new model for a physical phenomenon already addressed in xLPR to be easily substituted into the code. The new model might reflect new knowledge of the underlying physical process, new modeling techniques, an alternative methodology for use in comparative analyses and sensitivity studies, or a different but related physical process (e.g., irradiation-assisted stress-corrosion cracking versus PWSCC). While such "plug and play" flexibility is inherently desirable, its practical implementation actually imposes nontrivial limitations and burdens on the code that may act at cross-purposes to the goal itself. Modularity demands that the "object" be packaged in a manner that is consistent with a set of generic requirements defining all relevant parameters to allow it to "fit" in a particular location and properly interface with the system in which it will be inserted.

Within the xLPR framework, each computational module must "fit" in terms of its general analytical compatibility and interaction with the full suite of modules in the code. All inputs required by a new module must be available within the code, and specific outputs must be provided to downstream modules. However, the very nature of an alternative mathematical model reflecting new or different insights or a related but different physical process would suggest that the required inputs will likely differ, possibly significantly so, and in unpredictable directions. Accommodating new user inputs in such cases may be relatively straightforward, but dynamic inputs generated during a realization by other modules would, of course, be an entirely different matter. Anticipatory inclusion of coding to generate such intermediate data would be unrealistically burdensome and almost entirely speculative. Given the extensive, complex interaction between the modules in the code, attaining more than a modest degree of module interchangeability is not a practical goal. Furthermore, within the software QA environment to be implemented for the production code, any such changes to the released code must be accomplished in full compliance with the QA process, including validation and verification following addition of the new module. Therefore, incorporation and evaluation of alternative modules will primarily be restricted to the ongoing code development environment with consideration for formal inclusion of beneficial alternatives and enhancements within the normal versioning process.

6.3 Evaluation by the Center for Nuclear Waste Regulatory Analyses

6.3.1 Scope

CNWRA at Southwest Research Institute performed an independent evaluation [7] of the two framework codes against a set of criteria developed under the guidance of the Acceptance Criteria Task Group. This evaluation was structured to assess the following attributes of the two different framework codes:

1. Code Efficiency and Operational Convenience from End Users' Perspective

    a. Ease of modification of input data

    b. Ease of execution

    c. Ease of access of output data

2. Clarity and Readability from Independent Programmers' Perspectives

    a. Ability to make changes

    b. Internal documentation

    c. Compatibility

3. Adaptability and Flexibility

    a. Dummy module development

    b. Evaluate time to incorporate

    c. Evaluate existing framework functions

    d. Evaluate intermediate outputs

    e. Estimate time to come up to speed on each framework

4. Future Development Potential

    a. Preprocessing and post-processing

    b. Parallel and distributed processing

    c. Multiparty code development

    d. Third-party software implementation

    e. Code limitations

A summary of the conclusions of this assessment follows.

6.3.2 Assessment

xLPR-GoldSim

In general, xLPR-GoldSim and xLPR-SIAM have different limitations and strengths in regards to future development potential. xLPR-GoldSim is built on GoldSim which offers major strengths with respect to prompt model deployment, polished interfaces, graphic display, management of Monte Carlo data, limited background needed to read GoldSim model files, and quick learning time for model developers. GoldSim is a frame with numerous predefined functions that can be used in a "plug and play" approach. When specialized functions or approaches are needed, workarounds are possible to adapt existing GoldSim functions to perform different tasks. However, these workarounds generally compromise model clarity by the use of complex logics. Also, as GoldSim models grow in complexity, modifications can become cumbersome because adjustments need to be manually implemented in the fields of the GoldSim elements, element by element. The full evaluation report [7] noted some of these limitations of xLPR-GoldSim. However, most of these limitations are intrinsic to the GoldSim software and to resolve them would require changes in the software rather than changes in the xLPR-GoldSim model.

The double-nested loop structure of the xLPR Version 1.0 GoldSim code largely precluded use of the post-processing tools contained in GoldSim to perform computations on data stored in a model file. In the xLPR project, this limitation was overcome by manually exporting data stored in GoldSim model files. However, this process was inefficient and often took hours to complete. In addition, analyzing the exported data then requires analytical and plotting software other than GoldSim. To extend this example, the xLPR project will likely encounter limitations when additional models for failure of other components of the piping cooling system are developed using GoldSim. Chances are high that separate models would have to be programmed; each would generate "raw data," and external tools or software would be needed to analyze the GoldSim raw data to define a total system metric of risk. The process for a total system analysis is expected to be complex and, therefore, possibly limited to implementation by expert users.

However, discussions with GoldSim indicate that models separating uncertainties (aleatory and epistemic) have been successfully deployed without the nested-loop approach used for xLPR Version 1.0. If this approach can be successfully implemented for xLPR, the post-processing limitations experienced in Version 1.0 should be substantially mitigated. Furthermore, GoldSim code additions and enhancements funded through xLPR may also alleviate some of the other limitations and workarounds experienced with the Version 1.0 code.

In summary, models developed in GoldSim can be quickly deployed and are readable, and moderate effort is needed to initiate model maintenance. On the limiting side, developers must deal with GoldSim software constraints through workarounds as the model grows in size and complexity, the requirement for external tools to analyze the data seems unavoidable (for the double-nested loop structure), and data exporting is a potential bottleneck.

xLPR-SIAM

xLPR-SIAM, built on the SIAM-PFM (or SIAM for short) platform, is intended to include tools for probabilistic risk assessment and to be extensible to address different problems. Accordingly, in this evaluation, functionality was judged against that considered important in a frame for the development of stochastic models. Functionality expectations therefore included (1) sampling input parameters from distribution functions, (2) input data management from multiple realizations, (3) output data management (e.g., single value per realization, time series per realization, multi-value per realization, or multi-value per time step), and (4) plotting capabilities for inputs and outputs. To offer the intended frame adaptability, such functionality should be available systematically, so that by adapting blocks of code with minor changes (thus minimizing the need to write new, customized code), the functionality could be applied to different inputs and outputs and be available for the development of varied models. xLPR-SIAM scored high in input parameter sampling and input data management (areas (1) and (2) above). In dummy module trials, only a few lines of code were necessary to create a new parameter in an input tab in the xLPR-SIAM GUI with all of the pull down choices, create text log files, and enable graphic displays of probability density or cumulative probability plots. On the other hand, xLPR-SIAM did not score high in area (3), output data management.

Using the dummy module, changes were needed in several Python files and a FORTRAN code, plus compilation of the FORTRAN code, to capture a time series from multiple realizations in a text file. Based on this example case, the project team expected that different code changes are needed to export data, depending on which data are to be exported. Given this need for code customization, a deep level of expertise in Python and FORTRAN would be required to properly maintain a model, and an even deeper level of expertise would be needed for model development. This situation might be alleviated by defining standard data interfaces and standard data management functions systematically applicable to capture multiple realization data in text files or databases. Additionally, tutorials would need to be written for model developers. Modelers would then use predefined functions from SIAM libraries to build stochastic models. The learning of SIAM would be focused on learning the library functions with limited learning of Python; thus, SIAM would become accessible to a range of programmers. Currently, to implement a model following the xLPR-SIAM example, extensive knowledge of object-oriented programming, Python, FORTRAN, and Python libraries is needed.

With respect to area (4), plotting capabilities, xLPR-SIAM received a mixed score. A well structured systematic approach was identified for plotting data output. In general, only a few extra lines of source code were needed to make the data available in pull down menus in the SIAM GUI for the plots of output data. The identified plotting shortcomings are related to the lack of options to control the appearance of plots and the lack of plots for single-value outputs for a realization. Thus, effort is needed to develop the back-end of the SIAM frame (i.e., data management of multiple realization outputs and plotting of outputs) to make it a general frame for the construction of stochastic models, accessible to a range of programmers with diverse experience. It was also recommended that SIAM incorporate existing technologies for parallel processing, especially during the early stages of xLPR development.

All of the limitations noted for xLPR-SIAM can be addressed with extra coding effort. SIAM also offers the potential for scalability and developing an integrated unit for a total system risk analysis of piping cooling systems.

Cost Comparison

Thus, two options are envisioned for the xLPR future. GoldSim offers convenience at the cost of workarounds, the need for external tools, and investment in code changes by GoldSim. SIAM offers flexibility, scalability potential, and the possibility to develop integrated units for total risk assessments; however, extra investment is needed to build the frame to make it accessible to a range of programmers. To enhance the comparison of these two alternatives, the cost of use of GoldSim or SIAM by the NRC staff over the next 5 years was estimated. The unit of "cost" selected was the time of use. A longer time would be associated with a frame that is more expensive to use. A time estimate, as opposed to a dollar quantity, was preferred; as such information can be inferred more directly from the evaluation.

The following assumptions were adopted to estimate the time of use:

- SIAM is polished to improve accessibility to a broader range of programmers.

- Every year, a new module for addressing a piping system degradation attribute is implemented.

- Every year, one staff member is trained on the use of the frame to ensure continuity (to account for rotations and transfers).

- To translate between time and money, the following equivalence was used: 1 hour = $250.

The evaluators were asked to estimate the minimum and maximum time to undertake a task. To account for uncertainty, it was assumed that the actual task time could be any time between the minimum and maximum and would follow a uniform distribution.

In computing totals, costs assumed to be incurred only once in a 5-year period (e.g., GoldSim licenses are purchased only once; post-processing scripts to analyze GoldSim data are programmed only in the first year) are appropriately accounted for and other tasks are assumed to be performed every year. In the case of GoldSim, the comparison assumed that GTG would provide yearly training, three GoldSim licenses would be purchased, and license maintenance fees would be paid for access to recent versions of GoldSim. In the case of SIAM, the comparison accounted for the fact that SIAM libraries and dependencies are open source (i.e., are available at no monetary cost).

The main difference reflected in the use comparison is in the time needed to launch runs and obtain data to perform analyses. For example, in the case of GoldSim, up to 6 hours per run (modeler time, not runtime) was allowed for a modeler to set the run, organize the data, and execute post processing to derive meaningful results. In the case of SIAM, a maximum of 1 hour was allowed, under the assumption that the SIAM frame is developed to a mature state. On average, it is concluded that less time would be spent in using the SIAM frame, assuming SIAM is developed to a more mature state.

The time to develop SIAM to a mature state was also estimated, based on the recommendations from the evaluation, to obtain total estimates for the cost of the SIAM frame. When the SIAM development time is considered, the total time (cost) associated with the use of the SIAM frame exceeds the time (cost) associated with the use of the GoldSim frame by approximately 30 percent.

6.3.3 Conclusions

The evaluation concluded that for a set of defined assumptions over a 5-year evaluation time period, the SIAM frame would be approximately 30 percent more expensive than the GoldSim frame. However, considering only usage, the time necessary to exercise the SIAM frame was estimated to be lower than for the GoldSim frame, assuming that the SIAM frame is developed to an appropriately mature state.

SIAM is expected to be developed to a stage such that models would incorporate tools for post processing, thereby making the use of SIAM more convenient. In contrast, significant user intervention is expected in GoldSim models to analyze and interpret output data, unless GTG implements key new features. Such new features should include development of an approach to access and manipulate data stored in model files that does not require exporting to external text files.

Therefore, greater initial investment to sponsor the development and use of SIAM would reasonably be expected to result in a gain of flexibility and convenience. However, appropriate consideration should be given to the risk and cost of software development, especially in the absence of a commercial entity committed to long-term support and software maintenance, and frequent changes in hardware, operating systems, and third-party software.

Alternatively, use of GoldSim would likely cost less at the expense of more user intervention to execute models and analyze output data. However, while GTG is not likely to be able to resolve all identified limitations, the company has indicated a willingness and capability to add flexibility to address certain modeling limitations and data output restrictions.

6.4 Discussion

GTG was contacted to assess its flexibility and willingness to make changes to the GoldSim code that might better accommodate the needs of xLPR. User-specific versions of the GoldSim code are not maintained. Any code changes and enhancements that GTG agrees to make and xLPR agrees to fund would be incorporated into the next released version and would be available to all GoldSim licensees. While not all issues identified in the CNWRA report [7] as noteworthy limitations in GoldSim would be fully resolvable, GTG can likely address many of them to a satisfactory degree and at reasonable cost. Furthermore, with a substantial and diverse customer base, long-term code support appears assured.

As noted in the CNWRA report, necessary development within the SIAM platform can be readily accomplished to produce an xLPR code with a more streamlined user interface, particularly for output from the code. However, the upfront development costs, as well as ongoing code maintenance and support, represent a significant economic factor in the overall platform selection decision.

Informing final computational platform selection was the specific goal of developing parallel codes within the Pilot Study. The knowledge and experience gained relative to that goal have been invaluable in reaching this complex and challenging decision. Based on the structured comparison, cost analysis, and long-term prospects as described above, the xLPR project team has recommended that the future versions of xLPR be developed using the GoldSim commercial software as the computational framework. Section 8 of this report provides a summary of key recommendations regarding the direction for xLPR Version 2.0, including a detailed discussion of this framework selection decision for continued xLPR development.

7

LESSONS LEARNED AND KEY KNOWLEDGE GAPS

The xLPR Pilot Study exercised the process proposed for further xLPR development, but with the primary goal of evaluating and refining that process before undertaking the greater challenge associated with development of the first production version of the xLPR code. Over the course of this investigation, the xLPR project team developed an appreciation for the complexity of this problem. Successful completion of a comprehensive probabilistic fracture mechanics (PFM) code, even one initially rather narrowly focused on the dissimilar metal piping butt weld, encompasses a significant increase in complexity over the code addressed through the Pilot Study. Overall program success will depend on many factors, including the following:

- dedicated team members, collectively representing a full range of technical expertise consistent with the technical range of the xLPR problem, yet individually having a sufficient understanding of the basic computational process to efficiently support incorporation of elements into the overall model

- an enthusiasm for the project goals within the team and group leadership

- an efficient communication process within the project team

Throughout the approximately 18 months devoted to Pilot Study development and exercising of the code, the project team learned many lessons relevant to defining the path forward to production versions of the code. These lessons include those of a technical nature from module and framework development and implementation, as well as lessons related to organization and program management. During the project, various reports and tracking tools have captured these lessons. A summary discussion of the lessons learned follows.

7.1   Organizational Issues

As work progresses past the Pilot Study, it is imperative that the project organization, management, and decision making process be revisited, with specific attention to the following:

- The process for obtaining PIB review and approval was vague, cumbersome, and generally inconsistent with efficient project management and schedule adherence.

- While having "equal" co-leaders of each task group reflected the cooperative nature of the project, this arrangement also resulted in confused lines of responsibility for directing work activities within each group. There should be a single leader or "driver" at each level and each task area to "own" the task, push completion, enable information sharing, and ensure that documentation is completed on time.

7.2   Communication Issues

7.2.1 Direct Group Communication

Group interdependency is inherent in a program of this scope and magnitude, and consequently, constructive interactions and efficient communications within and between project groups are essential. The information needs, expectations, and priorities within and between groups must be efficiently and effectively communicated for the project to have any chance of success. For this reason, project planning and budgeting should allow for overlapping participation in group meetings and conference calls (i.e., by including additional man-hours and travel expenses).

7.2.2 Indirect Communication

The use of a Microsoft SharePoint site was an inherent element in the success of this project phase. With a geographically distributed team such as for xLPR, Internet-accessed collaborative Web tools for file sharing and archiving project documents are essential so that all participants have continuous access. However, while SharePoint worked very well for many Pilot Study written reports and related documentation, its limitations associated with handling very large files, particularly non-Microsoft-Office files, were also evident. These limitations must be addressed to support further xLPR development.

7.3   Framework Issues

7.3.1 Inputs and Outputs

For the Pilot Study, the input and output structures were unrefined because of the Pilot Study schedule and the need to demonstrate only the feasibility of the process. Since the input of data and the presentation of results are key usability attributes for any software tool, it is recommended that the development of much more efficient and flexible input and output structures for future versions of xLPR be emphasized.

7.3.2 Uncertainty Classification and Analysis

The classification of uncertainty is very important to understanding the overall uncertainty in the probability of rupture and should be considered at all levels in the development of a complex system. Knowing which variables control the rupture and which part of the uncertainty in those variables is epistemic and can be reduced will not only inform the users, but will also help direct future research in this area. In other words, if uncertainties are correctly characterized and prioritized, then xLPR can be used to prioritize research efforts and degradation management strategies to quantitatively maintain or improve safety.

However, uncertainty classification and quantification are not trivial. The Inputs Task Group has a major role in describing the uncertainty of each input, but that group must work with the Models Task Group and the Computational Task Group in order to understand exactly the context in which each input will be used. Ideally, uncertainty characterization should incorporate a feedback loop to ensure that results are reasonable and explainable, that no uncertain quantity has been incorrectly characterized, and that "inappropriate conservatism" is avoided.

In many cases, the data available that describe the uncertainty may be sparse, and the choice of the distribution to fit to that data may be arbitrary. Furthermore, the choice of which parameters are uncertain or constant, the classification of this uncertainty (aleatory or epistemic), or the selection of distribution type to represent uncertainty may greatly change the results of the analyses. For the input to Version 1.0, the uncertainties for some key input parameters were classified as exclusively either aleatory or epistemic. For the input to Version 2.0, partitioning the uncertainties for each key input parameter into both epistemic and aleatory components should be considered, since in most cases, both are present, although the epistemic is typically the larger of the two uncertainties. In addition, code development should retain the generic flexibility to allow the user to choose from a list of parameter classification types (constant, epistemic, aleatory) and distribution types (normal, lognormal, etc.) for most, if not all, inputs.

7.3.3 *Improved Sampling Techniques*

The calculation of low-probability events for a complex system with a variety of random inputs can be extremely difficult using standard sampling techniques. Not only is an extreme number of standard realizations required, the data storage capacity for running such analyses is prohibitive. Since many of the low-probability events occur when the tails of the input distributions are controlling the event, improved sampling techniques, such as importance sampling, stratified sampling, or adaptive sampling, are required in order to produce acceptable results in a reasonable time, while remaining within reasonable data storage capacity limits.

The Pilot Study has demonstrated that low-probability events (less than 10^{-6}) can be calculated with relative ease when the correct variables are sampled using improved techniques. However, multiple sensitivity analyses likely must be conducted before it is evident which variables are important to the output of interest. The use of improved sampling techniques is conditional to a good understanding of the system and the effect of each sampled parameter on the outputs of interest. The choices of the distribution used and, to a lesser extent, the parameters selected, are mainly based on user experience. A bad selection may focus the analysis on the wrong area (either an area without interest, or an area of such low probability of occurrence that it will not affect the final result), which could make the sampling results unreliable or, even worse, misleading.

Therefore, improved sampling methods should be further developed and incorporated into future versions of the xLPR code. In addition, other optimization and reliability methods have been developed to handle this issue and should be studied for possible use in future versions of the xLPR code. Finally, processes and procedures for identifying the key variables that need to be sampled using the improved techniques should be emphasized.

7.3.4 *Data Storage and Handling*

For the purposes of the Pilot Study, all of the data from each realization generated from the probabilistic runs were stored to maintain maximum flexibility to extract any results desired by post processing the data instead of re-running the code. However, this leads to large result files that are extremely difficult to handle, and in many cases, this prevented run completion because of hard drive storage capacity limits despite the use of compressed data formats for storing results. In addition, outputting the results into usable text format was a time-consuming process that, in some cases, took longer than the code run itself. Therefore, for future versions of the xLPR code, the data storage and handling process should be revisited to streamline the amount of data saved and to output only the data necessary for understanding the case results and sensitivities being investigated.

7.3.5 Post Processing

In the development of the xLPR Version 1.0 code, most of the development time was spent creating a framework that performed the calculations required to determine whether rupture occurs in a given realization. Not until near the end of the Pilot Study did the focus shift to development of post-processing tools needed for the calculation of rupture probabilities from the cumulative results, and evaluating the effects of uncertainty characterizations (aleatory versus epistemic), leak detection, and inspection on the output. Additional post-processing software will also likely be needed for parameter sensitivity analyses when non-monotonic influences between inputs and outputs are present. Therefore, for future versions of the xLPR code, it is recommended that sufficient time be allotted to the development of post-processing software. It is imperative that correct and easy-to-use software be available to post process the large dataset that is developed from this complex probabilistic code.

7.4 Models Issues

7.4.1 Expertise

Overall, the expertise applied to the Pilot Study tasks was appropriate and sufficient, but for the Models Group in particular, some subgroups may have lacked sufficient breadth of expertise to most effectively evaluate and recommend the most appropriate models for use. The issues of staffing in the Models Group stemmed from the availability of funds and resources, as well as the impact of other work priorities on group members' time, not the ability to find the appropriate staff. Priorities and other commitments limited many model experts' available time.

7.4.2 Modeling Scope

Certain simplifying assumptions within the modeling area were made for the Pilot Study that were consistent with its limited scope, but should be reconsidered for subsequent xLPR development. These model-based issues include the following:

- Manufacturing defects and fatigue initiation and growth were ignored in the Pilot Study. Inclusion of these items, with appropriate concern for their associated complexity, should be considered for future versions so that their effect on the calculated rupture probabilities can be evaluated.

- The load module should be updated to include a more realistic weld residual stress model, including its variation around the circumference, and transient definitions.

- Considering only circumferential cracks may over predict the rupture probabilities. The higher leakage probabilities attributable to axial cracks, could lead to early detection followed by repair or mitigation, and as a result reduce the rupture probability.

- The assumption of idealized flaw shapes and simplistic transitions from a surface crack to a through-wall crack may cause an overestimate of the leak rate and should also be reevaluated.

- More realistic surface crack stability, inspection, and mitigation models should be considered for making best estimate predictions of their effects.

Because many technical areas of interest are closely related to this project, it is vital that the scope be well defined and observed. However, it will also be necessary to have a defined process for critically evaluating "enhancement opportunities" to determine whether to accept the added scope, identify it for consideration in a future version, or drop it from consideration.

7.5   Software Quality Assurance and Configuration Management

CM and software QA (SQA) are necessary and beneficial, yet also have an impact on the software development scope and schedule. SQA consists of a systematic and documented practice of monitoring the software and model development processes and methods used to ensure quality. SQA encompasses the entire software life cycle, which includes processes such as requirements definition, software design, coding, source code control, code reviews, change management, CM, testing, release management, and product integration. SQA is organized into goals, commitments, abilities, activities, measurements, and verification and validation and typically follows a process consistent within the industry (e.g., International Organization for Standardization (ISO)-9001 [10], American Society of Mechanical Engineers (ASME)-NQA-1-2008 [11], ASME VV 20-2009 [12]), regardless of the application. Model development follows a similar process, which incorporates the fundamental aspects of QA, including version control, reviews, change management, testing, CM, and release management. CM is the process that focuses on demonstration, documentation, and control of the steps taken and the products developed under a QA program. A robust CM system includes both electronic and programmatic controls that are linked to a well-defined QA program. The link between the CM and QA program usually takes the form of guidelines or a CM plan, which provides the roadmap between the required QA steps and methods and the CM system that maintains the configuration control.

While a relatively robust CM program was implemented within the Pilot Study, the overarching SQA element was not. However, use of the xLPR code and acceptance of its results for safety related risk analyses dictate that code development proceed under the auspices of a suitable QA program consistent with the expectations contained in NUREG/BR-0167, "Software Quality Assurance Program and Guidelines," issued February 1993 [13]. Key actions include development of the following:

- An xLPR program QA plan and controls as the essential first step in the continuing development process.

- A transparent and traceable CM system that will cover the xLPR code life cycle.

- A well-written, unambiguous software requirements document defining the detailed scope for future xLPR versions.

8
RECOMMENDATIONS FOR VERSION 2.0

The xLPR Pilot Study is the culmination of the initial development of a sophisticated, thorough, and quality-assured software tool for assessment of probabilistic failure. The Pilot Study project team developed and exercised the initial version of this code on a limited scope problem to assess the structure and feasibility of the management and technical approach. The main goal of the Pilot Study was threefold:

(1) Assess the proposed management structure's ability to support cooperative and efficient code development and implementation.

(2) Assess the feasibility of developing a modular-based PFM computer code that can calculate the probability of rupture for a reactor coolant nozzle weld while properly accounting for the problem uncertainties.

(3) Determine the appropriate probabilistic framework for constructing the modular-based PFM code.

The Pilot Study effort encompassed the code development work, exercising the project management structure, implementing the Pilot Study problem statement, and detailed analysis of the results. Through this effort, the xLPR project team demonstrated that it is feasible to develop a modular-based computer code for the determination of probability of rupture for LBB-approved piping systems.

The following sections address the overall conclusions from the Pilot Study and recommendations for further xLPR development beginning with Version 2.0, with respect to each of the three Pilot Study goals.

8.1   Project Management Structure

Through a cooperative effort between the NRC and EPRI, guided by an addendum [1] to the ongoing memorandum of understanding between the two organizations, a project management structure was developed with balanced NRC and industry representation for the xLPR Pilot Study, consisting of four topical technical task groups (Computational, Models, Inputs, and Acceptance Criteria) coordinated by the overarching PIB. The NRC and EPRI staffed the four technical task groups with a variety of experts in solid mechanics, materials engineering, PFM, weld residual stress, fluid mechanics, computer programming, CM, and uncertainty treatment and propagation. The choice of the technical experts and their assignment to task groups proved highly successful through the completion of the Pilot Study problem. The role of the PIB was to provide overall guidance, coordinate the work, and make programmatic decisions on an as-required basis. While the technical team experts covered technical issues, and the task group leads coordinated communication within their own task groups, attaining effective communication between teams and with the PIB proved more challenging. Early in the program, limited overlap of personnel between the task groups led to missed action items or miscommunication between the teams and created frustration. Although the PIB was ostensibly responsible for this communication coordination role, distributing this important responsibility across a committee of 12 proved completely ineffective. The task group leads responded by attending each other's meetings to achieve the necessary coordination of task group work, which greatly increased the communication between groups.

While the task groups found efficient ways to facilitate progress, communication and coordination through the PIB was not as effective. The PIB consisted of two members from each task group and a management and at-large representative from both the NRC and industry. This group was instrumental in providing high-level project direction, but of these 12 members, no one was identified as the leader, and no real process for PIB review and approval was ever developed. In addition, the relatively compressed schedule for the Pilot Study necessitated quick decisions by the task groups and rendered routine approval of such decisions by the PIB impractical. However, even without the input of the PIB, the project team successfully completed the Pilot Study and produced results that demonstrated the feasibility of the approach. This achievement is attributable to the leadership of the technical task groups.

Therefore, the overarching structure of the PIB is not required for success of this effort. In fact, a panel of 12 engineers and managers will almost certainly be ineffective in providing routine project management. As the program progresses, the project team recommends a restructuring of the project organization. Clearly, the advantage of the PIB was its ability to provide high-level guidance and overall program review. Therefore, the project team recommends that the PIB be restructured as an advisory/review committee and that the decision-making authority be assigned to a code development leader responsible for facilitating communication between task groups and guiding and focusing the task groups toward the final goal.

8.2 Modular Code Feasibility

Through the efforts documented in [5], the project team successfully demonstrated that it is feasible to develop a modular-based computer code for the determination of primary piping rupture probabilities. In fact, the team developed two separate framework codes using a common set of deterministic modules that were interchangeable with modules with similar inputs and outputs. The project team conducted a set of analyses using each framework that demonstrated the ability to calculate the probability of rupture under operating conditions taking into account PWSCC growth rates, inspections, and leak detection limits. In addition, the problem uncertainties were quantified and propagated throughout the problem such that a distribution of rupture probabilities was generated for each Pilot Study problem case. The project team concluded that improved sampling techniques are required for low rupture probability calculations and that uncertainty classification and quantification are not trivial and should be considered at all levels of development for this complex system.

In this complex problem where many factors influence the desired results, the project team could not develop the framework independently from the rest of the analysis, but instead considered it as an integrated part of the whole project. Furthermore, the framework development cannot be considered as a simple "plugging in" of modules in a probabilistic loop. On many occasions, as the team developed and validated the framework so that the flow between modules was appropriate, it was necessary to revise the framework logic or a preceding module to correctly capture the phenomenon considered or to appropriately represent the response in a downstream model. Adding new modules or models where the inputs and outputs vary greatly from the implemented modules therefore will always require some modification to the framework, and it is necessary that the project team understand how a new module affects the downstream modules and the overall probabilistic flow.

The project team will develop the final xLPR code to meet quality standards (e.g., ASME-NQA-1-2008 [11]). However, the team did not perform this task for the Pilot Study because feasibility determination was the primary goal. Therefore, the project team focused on the development of a comprehensive CM process that would generally satisfy QA requirements. The project team concluded that this process was sufficient for the feasibility study, but lacks the overarching structure of a QA program. SQA encompasses the entire software development process, including processes such as requirements definition, software design, coding, source code control, code reviews, change management, CM, testing, release management, and product integration. The link between the CM and the QA program usually takes the form of guidelines or a CM plan, which provides the roadmap between the required QA steps and methods and the CM system that maintains the configuration control. Even though the development of the Version 1.0 code followed a defined CM structure, it was not linked to a QA program, which resulted in some difficulty and disconnection in the areas of validation, life-cycle definition, and management coordination. For future versions of the code, the project team recommends that a single, project-wide QA program be developed, or the several QA systems located in the various xLPR organizations be integrated into a complete QA program for xLPR.

8.3 Computational Framework

Within the xLPR Pilot Study, the project team successfully implemented two unique framework codes to investigate the advantages and disadvantages of two approaches: (1) use of available commercial software and (2) use of open source code. The project team used the commercial software GoldSim to investigate the commercial software approach for the xLPR model, and used the open source code SIAM-PFM to demonstrate the open source approach. Within these frameworks, the project team implemented essentially the same program flow and deterministic modules through a detailed CM program. After completing a verification process, the project team exercised each code through a set of Pilot Study sample problems and compared these results in [5]. Even though it identified and documented some slight differences between the results, the project team found that the comparison between results was favorable and differences were explainable.

An independent contractor conducted a comparison of the two framework codes [7] and concluded that each framework has its own set of strengths and limitations, none of which greatly promotes one framework over the other. GoldSim offers prompt model deployment, polished interfaces, graphic display, management of Monte Carlo data, easy to understand GoldSim model files, and quick learning time for model developers but at the cost of workarounds and the possible need for framework enhancements or external tools. SIAM-PFM offers flexibility, scalability potential, and the possibility of developing integrated units for total risk assessments involving multiple "components"; however, extra investment is needed to further develop the framework to not only make it accessible to a range of programmers, but also to bring it to the level of sophistication embodied in the GoldSim framework.

To aid in the framework selection process, the independent contractor conducted a cost analysis with the following assumptions:

- use of either the GoldSim or SIAM framework by the NRC staff, over the next 5 years

- further development of the SIAM framework to provide features similar to GoldSim

- inclusion of all applicable GoldSim licensing and training costs

- annual implementation of a new module for failure of piping systems

- annual training of one NRC staff member on the use of the framework to ensure continuity (to account for rotations and transfers)

The contractor concluded that the SIAM code would be about 30 percent more expensive than GoldSim over this 5-year period. Most of this cost difference was attributed to the second assumption above.

The long-term prospect of software maintenance and support is essential to ensure that the life cycle of this computer code extends beyond the range of industry and regulatory use assumed above. A commercial software package is attractive because the revenue needed for the development and maintenance of the software is supplied by a variety of customers and not reliant on a single funding vehicle. In addition, a commercial software development company has a management and overhead structure optimized for the long-term maintenance of software that engineering contract companies or national laboratories lack. Finally, the fear of bankruptcy for commercially driven software companies may be eased by the placement of the source code into escrow as a contract provision, which ensures delivery of the complete source code if the company is dissolved or goes bankrupt.

The selection of an appropriate framework for the further development of xLPR should be based on not only the technical capabilities, but also the cost impact and longevity potential of the software. For the case of xLPR, since the project team demonstrated that both GoldSim and SIAM possess similar technical capability potential, cost and longevity will drive the selection process. For open source software, the project team can develop appropriate coding features, implemented and maintained exactly as required by the project, given sufficient time and resources. However, in today's economy, resources are limited. Commercial software companies are driven to be as diverse in funding opportunities as possible in order to survive. This can translate to large cost savings to the xLPR program in that other GoldSim clients may fund modifications to GoldSim that benefit xLPR, or GoldSim may fund proposed modifications that would benefit all clients. For project-specific modifications, the cost of directly funding GoldSim to make these changes is far less than directly funding a national laboratory to make changes to a noncommercial code. In addition, with the source code escrow process, longevity of the commercial software is ensured. Therefore, based on the comparison, cost analysis, and long-term prospects, the xLPR project team recommends that future versions of xLPR be developed using the GoldSim commercial software as the computational framework.

9 REFERENCES

1. B. Sheron and D. Modeen, "Addendum to Memorandum of Understanding between U.S. Nuclear Regulatory Commission and Electric Power Research Institute on Cooperative Nuclear Safety Research—Extremely Low Probability of Rupture (xLPR)," October 2009, ADAMS Accession No. ML092290118.

2. P.D. Mattie, C.J. Sallaberry, J.C. Helton, and D.A. Kalinich, "Development, Analysis, and Evaluation of a Commercial Software Framework for the Study of Extremely Low Probability of Rupture (xLPR) Events at Nuclear Power Plants," SAND2010-8480, Letter report to the Office of Nuclear Regulatory Research, U.S. Nuclear Regulatory Commission, Sandia National Laboratories, Albuquerque, NM, December 2010, ADAMS Accession No. ML110700019.

3. H.B. Klasky, P.T. Williams, S. Yin, and B.R. Bass, "SIAM-xLPR Version 1.0 Framework Report," ORNL/NRC/LTR-248, Letter report to the Office of Nuclear Regulatory Research, U.S. Nuclear Regulatory Commission, Oak Ridge National Laboratory, Oak Ridge, TN, September 2010, ADAMS Accession No. ML110700026.

4. Electric Power Research Institute, "Models and Inputs Developed for Use in the xLPR Pilot Study," PID 1022528, Palo Alto, CA, 2011.

5. D. Rudland, "xLPR Version 1.0 Report—Technical Basis and Pilot Study Problem Results," U.S. Nuclear Regulatory Commission, Washington, DC, February 2011, ADAMS Accession No. ML110660292.

6. Electric Power Research Institute, "Materials Reliability Program: Advanced FEA Evaluation of Growth of Postulated Circumferential PWSCC Flaws in Pressurizer Nozzle Dissimilar Metal Welds" (MRP-216), Rev. 1, PID 1015383, Palo Alto, CA, 2007.

7. Pensado, O., et al., "Assessment of Capabilities of Extremely Low Probability of Rupture (xLPR) Software—GoldSim and SIAM Version 1.0," Center for Nuclear Waste Regulatory Analyses, May 2011, San Antonio, TX, ADAMS Accession No. ML111510924.

8. P.D. Mattie, D.A. Kalinich, and C.J. Sallaberry, "U.S. Nuclear Regulatory Commission Extremely Low Probability of Rupture Pilot Study: xLPR Framework Model User's Guide," SAND2010-7131, Sandia National Laboratories, Albuquerque, NM, November 2010, ADAMS Accession No. ML110700017.

9. H.B. Klasky, P.T. Williams, S. Yin, and B.R. Bass, "Structural Integrity Assessments Modular—Probabilistic Fracture Mechanics (SIAM-PFM): User's Guide for xLPR," ORNL/NRC/LTR-247, Oak Ridge National Laboratory, Oak Ridge, TN, September 2010, ADAMS Accession No. ML110700023.

10. International Organization for Standardization, ISO 9001:2008, "Quality Management Systems—Requirements," Geneva, Switzerland.

Insert Appropriate Auto Text License Entry. If license is copyright, please delete

Error! No text of specified style in document.

11. American Society of Mechanical Engineers, ASME NQA-1-2008, "Quality Assurance Requirements for Nuclear Facility Applications," and ASME NQA-1a-2009, Addenda to ASME NQA-1-2008, "Quality Assurance Requirements for Nuclear Facility Applications," New York, NY.

12. American Society of Mechanical Engineers, ASME VV 20-2009, "Standard for Verification and Validation in Computational Fluid Dynamics and Heat Transfer," New York, NY.

13. U.S. Nuclear Regulatory Commission, NUREG/BR-0167, "Software Quality Assurance Program and Guidelines," Washington, DC, February 1993.

NRC FORM 335
(12-2010)
NRCMD 3.7

U.S. NUCLEAR REGULATORY COMMISSION

BIBLIOGRAPHIC DATA SHEET

(See instructions on the reverse)

1. REPORT NUMBER (Assigned by NRC, Add Vol., Supp., Rev., and Addendum Numbers, If any.)
NUREG-2110

2. TITLE AND SUBTITLE	3. DATE REPORT PUBLISHED	
xLPR Pilot Study Report	MONTH	YEAR
	May	2012
	4. FIN OR GRANT NUMBER	

5. AUTHOR(S)	6. TYPE OF REPORT
David Rudland, US NRC	Technical
Craig Harrington, EPRI	7. PERIOD COVERED (Inclusive Dates)
	NA

8. PERFORMING ORGANIZATION - NAME AND ADDRESS (If NRC, provide Division, Office or Region, U. S. Nuclear Regulatory Commission, and mailing address; if contractor, provide name and mailing address.)

U.S. Nuclear Regulatory Commission
Office of Nuclear Regulatory Research
Washington, DC 20555

Electric Power Research Institute
3420 Hillview Avenue
Palo Alto, CA 94304

9. SPONSORING ORGANIZATION - NAME AND ADDRESS (If NRC, type "Same as above", if contractor, provide NRC Division, Office or Region, U. S. Nuclear Regulatory Commission, and mailing address.)

Same as above

10. SUPPLEMENTARY NOTES

11. ABSTRACT (200 words or less)

Under the auspices of an addendum to the memorandum of understanding between the Electric Power Research Institute and the U.S. Nuclear Regulatory Commission's Office of Nuclear Regulatory Research for cooperative research, a pilot study has been completed to evaluate the feasibility of developing a fully probabilistic, fracture-mechanics-based computational tool to evaluate the rupture probability of reactor coolant piping. This project, known as xLPR for Extremely Low Probability of Rupture, is initially focused on evaluating pipe rupture probabilities within Alloy 82/182 dissimilar metal welds located in lines licensed for leak before break (LBB) as allowed under General Design Criterion 4.

This report summarizes the results of that pilot study. The xLPR Pilot Study team demonstrated that it is feasible to develop a modular-based computer code for the determination of probability of rupture for LBB-approved piping systems. Furthermore, while the organization established to manage the project and accomplish the technical work of the Pilot Study successfully met that challenge, it identified important improvement opportunities that will be addressed as the project moves forward. Finally, the project team selected a commercially licensed simulation framework code to be used in future versions of xLPR.

12. KEY WORDS/DESCRIPTORS (List words or phrases that will assist researchers in locating the report.)	13. AVAILABILITY STATEMENT
xLPR	unlimited
LBB	14. SECURITY CLASSIFICATION
Pipe rupture	*(This Page)* unclassified
	(This Report) unclassified
	15. NUMBER OF PAGES
	16. PRICE

DISCLAIMER OF WARRANTIES AND LIMITATION OF LIABILITIES

THIS DOCUMENT WAS PREPARED BY THE ORGANIZATION(S) NAMED BELOW AS AN ACCOUNT OF WORK SPONSORED OR COSPONSORED BY THE ELECTRIC POWER RESEARCH INSTITUTE, INC. (EPRI). NEITHER EPRI, ANY MEMBER OF EPRI, ANY COSPONSOR, THE ORGANIZATION(S) BELOW, NOR ANY PERSON ACTING ON BEHALF OF ANY OF THEM:

(A) MAKES ANY WARRANTY OR REPRESENTATION WHATSOEVER, EXPRESS OR IMPLIED, (I) WITH RESPECT TO THE USE OF ANY INFORMATION, APPARATUS, METHOD, PROCESS, OR SIMILAR ITEM DISCLOSED IN THIS DOCUMENT, INCLUDING MERCHANTABILITY AND FITNESS FOR A PARTICULAR PURPOSE, OR (II) THAT SUCH USE DOES NOT INFRINGE ON OR INTERFERE WITH PRIVATELY OWNED RIGHTS, INCLUDING ANY PARTY'S INTELLECTUAL PROPERTY, OR (III) THAT THIS DOCUMENT IS SUITABLE TO ANY PARTICULAR USER'S CIRCUMSTANCE; OR

(B) ASSUMES RESPONSIBILITY FOR ANY DAMAGES OR OTHER LIABILITY WHATSOEVER (INCLUDING ANY CONSEQUENTIAL DAMAGES, EVEN IF EPRI OR ANY EPRI REPRESENTATIVE HAS BEEN ADVISED OF THE POSSIBILITY OF SUCH DAMAGES) RESULTING FROM YOUR SELECTION OR USE OF THIS DOCUMENT OR ANY INFORMATION, APPARATUS, METHOD, PROCESS, OR SIMILAR ITEM DISCLOSED IN THIS DOCUMENT.

NRC DISCLAIMER

THE STATEMENTS, FINDINGS, CONCLUSIONS AND RECOMMENDATIONS ARE THOSE OF THE AUTHOR(S) AND DO NOT NECESSARILY REFLECT THE VIEW OF THE U.S. NUCLEAR REGULATORY COMMISSION.

THE FOLLOWING ORGANIZATIONS PREPARED THIS REPORT:

Electric Power Research Institute (EPRI)

U.S. Nuclear Regulatory Commission (NRC), Office of Nuclear Regulatory Research (RES)

THE TECHNICAL CONTENTS OF THIS DOCUMENT WERE **NOT** PREPARED IN ACCORDANCE WITH THE EPRI NUCLEAR QUALITY ASSURANCE PROGRAM MANUAL THAT FULFILLS THE REQUIREMENTS OF 10 CFR PART 50, APPENDIX B; AND 10 CFR PART 21; ANSI N45.2-1977; AND/OR THE INTENT OF ISO-9001 (1994). USE OF THE CONTENTS OF THIS DOCUMENT IN NUCLEAR SAFETY OR NUCLEAR QUALITY APPLICATIONS REQUIRES ADDITIONAL ACTIONS BY USER PURSUANT TO THEIR INTERNAL PROCEDURES.

NOTE

For further information about EPRI, call the EPRI Customer Assistance Center at 800.313.3774 or e-mail askepri@epri.com.

Electric Power Research Institute, EPRI, and TOGETHER...SHAPING THE FUTURE OF ELECTRICITY are registered service marks of the Electric Power Research Institute, Inc.

NUREG-2110

xLPR Pilot Study Report

May 2012

www.ingramcontent.com/pod-product-compliance
Lightning Source LLC
Chambersburg PA
CBHW081846170526
45167CB00007B/2912